Lecture Notes in Computer Science 8330

Commenced Publication in 1973
Founding and Former Series Editors:
Gerhard Goos, Juris Hartmanis, and Jan van Leeuwen

T0240523

Oscar Camara   Tommaso Mansi
Mihaela Pop   Kawal Rhode
Maxime Sermesant   Alistair Young (Eds.)

Statistical Atlases and Computational Models of the Heart

Imaging and Modelling Challenges

4th International Workshop, STACOM 2013
Held in Conjunction with MICCAI 2013
Nagoya, Japan, September 26, 2013
Revised Selected Papers

 Springer

Volume Editors

Oscar Camara
Universitat Pompeu Fabra, Barcelona, Spain
E-mail: oscar.camara@upf.edu

Tommaso Mansi
Siemens Corporation, Corporate Technology, Princeton, NJ, USA
E-mail: tommaso.mansi@siemens.com

Mihaela Pop
University of Toronto, ON, Canada
E-mail: mihaela.pop@utoronto.ca

Kawal Rhode
King's College London, UK
E-mail: kawal.rhode@kcl.ac.uk

Maxime Sermesant
Inria, Sophia Antipolis, France
E-mail: maxime.sermesant@inria.fr

Alistair Young
University of Auckland, New Zealand
E-mail: a.young@auckland.ac.nz

ISSN 0302-9743   e-ISSN 1611-3349
ISBN 978-3-642-54267-1   e-ISBN 978-3-642-54268-8
DOI 10.1007/978-3-642-54268-8
Springer Heidelberg New York Dordrecht London

Library of Congress Control Number: 2014930560

CR Subject Classification (1998): J.3, H.4, H.5.1-2, H.1.2, I.2.10

LNCS Sublibrary: SL 6 – Image Processing, Computer Vision, Pattern Recognition, and Graphics

Typesetting: Camera-ready by author, data conversion by Scientific Publishing Services, Chennai, India

Printed on acid-free paper

Springer is part of Springer Science+Business Media (www.springer.com)

Preface

Recently, there has been considerable progress in cardiac image analysis techniques, cardiac atlases, and computational models, which can integrate data from large-scale databases of heart shape, function, and physiology. Integrative models of cardiac function are important for understanding disease, evaluating treatment, and planning intervention. However, significant clinical translation of these tools is constrained by the lack of complete and rigorous technical and clinical validation, as well as benchmarking of the developed tools. For doing so, common and available ground-truth data capturing generic knowledge of the healthy and pathological heart are required. This knowledge can be acquired through the building of statistical models of the heart. Several efforts are now devoted to providing Web-accessible structural and functional atlases of the normal and pathological heart for clinical, research, and educational purposes. We believe all these approaches will only be effectively developed through collaboration across the full research scope of the imaging and modelling communities.

STACOM 2013 was held in conjunction with the MICCAI 2013 conference (Nagoya, Japan), and followed the last three editions: STACOM 2012 (Nice, France), STACOM 2011 (Toronto, Canada), and STACOM 2010 (2010, Beijing, China). STACOM 2013 provided a forum for the discussion of the latest developments in the areas of statistical atlases and computational imaging and modelling of the heart. The topics of the workshop included: cardiac image processing, atlas construction, statistical modelling of cardiac function across different patient populations, cardiac mapping, cardiac computational physiology, model customization, atlas-based functional analysis, ontological schemata for data and results, integrated functional and structural analyses, as well as the preclinical and clinical applicability of these methods. STACOM 2013 drew more than 40 submissions from around the world, with 31 papers finally accepted in the workshop and invited to be published in this *Lecture Notes in Computer Science* volume (Springer). Besides regular contributions on state-of-the-art cardiac image analysis techniques, atlases, and computational models that integrate data from large-scale databases of heart shape, function, and physiology, additional efforts of this year's workshop focused on two imaging and modelling challenges, described below.

CFD Challenge – The objective of the STACOM 2013 CFD challenge was to investigate the predictive power of CFD tools in terms of pressure gradient through an aortic coarctation at stress. Challengers were given preoperative data of one patient at rest and stress, with pregenerated geometrical model, MR-derived flow splits and invasive pressure measurements. They were then asked to predict the pressure drop through the coarctation at rest and stress using the computational tool of their choice. The papers gathered in these proceedings reported the methodology and the results obtained by each participant. During the workshop, variations in methodology and parameter sensitivity were discussed. Results were compared with "ground-truth" invasive measurements acquired via pressure wires. This year, the challenge attracted 11 groups worldwide, among them eight submitted final results. A journal paper summarizing the outcomes of the STACOM 2012 and STACOM 2013 CFD challenges is in preparation.

Left Atrium Segmentation Challenge – The left atrium is clinically important for the management of atrial fibrillation in patients. MRI and CT are commonly used for imaging this structure. Segmentation can be used to generate anatomical models that can be employed in guided treatment and also more recently for cardiac biophysical modelling. A total of 30 MRI and 30 CT datasets were provided to participants for segmentation of the endocardial boundary and the pulmonary veins up to the first branch point. Initially, ten datasets for each modality were provided with expert manual segmentations for algorithm training. The other 20 datasets per modality were used for evaluation. The datasets were provided by King's College London and Philips Research Hamburg. The challenge raised interest from 12 research groups worldwide. From them, five groups submitted final results and papers that were accepted to be presented at the workshop. Additionally, a collaborative article describing the unified benchmarking framework implemented for the challenge was also included by the LASC organizers and was presented during the workshop.

We hope that the results obtained by these two challenges, together with all regular paper contributions, will act to accelerate progress in the important areas of heart function and structure analysis.

September 2013

Oscar Camara
Tommaso Mansi
Mihaela Pop
Kawal Rhode
Maxime Sermesant
Alistair Young

Organization

We would like to thank the organizers, additional reviewers, senior advisors, and all participants for their time and effort in making STACOM 2013 a successful event.

Chairs

Oscar Camara   Universitat Pompeu Fabra, Barcelona, Spain
Tommaso Mansi   Siemens Corporation, Corporate Technology, Imaging and Computer Vision, Princeton, NJ, USA
Mihaela Pop   Sunnybrook Research Institute, University of Toronto, ON, Canada
Kawal Rhode   King's College London, UK
Maxime Sermesant   Inria, Sophia Antipolis, France
Alistair Young   University of Auckland, New Zealand

Challenges – Organizing Teams

CFD Challenge

T. Mansi
A. Figueroa
N. Wilson
P. Sharma

Left Atrial Segmentation Challenge

K. Rhode
C. Tobon-Gomez
J. Peters
J. Weese
R. Karim

Website

Avan Suinesiaputra   University of Auckland, New Zealand

Senior Advisors

Ali Kamen   Siemens Corporation, Corporate Technology, Princeton, NJ, USA
Boudewijn Lelieveldt   Leiden University Medical Center, The Netherlands
Graham Wright   University of Toronto, Sunnybrook Research Institute, Canada

Additional Reviewers

We would like to acknowledge the following reviewers who, in addition to the organizers of the workshop and challenges, provided scientific feedback to the participants on their papers:

Bart Bijnens
Constantine Butakoff
Rocío Cabrera Lozoya
Rubén Cardenes
Brett Cowan
Mathieu de Craene
Nicolas Duchateau
Alberto Gómez
Stéphanie Marchesseau
Kristin McLeod
Pau Medrano-Gracia
Martyn Nash
Gemma Piella
Avan Suinesiaputra
Federico Sukno
Nicolas Toussaint
Robert Xu
Vicky Wang

Table of Contents

Left Atrial Segmentation Challenge

Left Atrial Segmentation Challenge: A Unified Benchmarking
Framework 1
    Catalina Tobon-Gomez, Jochen Peters, Juergen Weese,
    Karen Pinto, Rashed Karim, Tobias Schaeffter, Reza Razavi, and
    Kawal S. Rhode

Automatic Segmentation of the Left Atrium on CT Images 14
    Abdelaziz Daoudi, Saïd Mahmoudi, and Mohammed Amine Chikh

Multi-atlas-Based Segmentation of the Left Atrium and Pulmonary
Veins 24
    Zulma Sandoval, Julián Betancur, and Jean-Louis Dillenseger

Model-Based Segmentation of the Left Atrium in CT and MRI Scans ... 31
    Birgit Stender, Oliver Blanck, Bo Wang, and Alexander Schlaefer

Toward an Automatic Left Atrium Localization Based on Shape
Descriptors and Prior Knowledge 42
    Mohammed Ammar, Saïd Mahmoudi,
    Mohammed Amine Chikh, and Amine Abbou

Decision Forests for Segmentation of the Left Atrium from 3D MRI 49
    Ján Margeta, Kristin McLeod, Antonio Criminisi, and
    Nicholas Ayache

CFD Challenge

Multiscale Study on Hemodynamics in Patient-Specific Thoracic Aortic
Coarctation 57
    Xi Zhao, Youjun Liu, Jinli Ding, Mingzi Zhang, Wenyu Fu,
    Fan Bai, Xiaochen Ren, and Aike Qiao

Hemodynamic in Aortic Coarctation Using MRI-Based Inflow
Condition 65
    Jens Schaller, Leonid Goubergrits, Pavlo Yevtushenko,
    Ulrich Kertzscher, Eugénie Riesenkampff, and Titus Kuehne

Sensitivity Analysis of the Boundary Conditions in Simulations
of the Flow in an Aortic Coarctation under Rest and Stress
Conditions 74
    Salvatore Cito, Jordi Pallarés, and Anton Vernet

Patient-Specific Hemodynamic Evaluation of an Aortic Coarctation
under Rest and Stress Conditions 83
    *Priti G. Albal, Tyson A. Montidoro, Onur Dur, and
    Prahlad G. Menon*

CFD Challenge: Predicting Patient-Specific Hemodynamics at Rest
and Stress through an Aortic Coarctation 94
    *Christof Karmonik, Alistair Brown, Kristian Debus,
    Jean Bismuth, and Alain B. Lumsden*

A Multiscale Filtering-Based Parameter Estimation Method
for Patient-Specific Coarctation Simulations in Rest and Exercise 102
    *Sanjay Pant, Benoit Fabrèges, Jean-Frédéric Gerbeau, and
    Irene E. Vignon-Clementel*

A Finite Element CFD Simulation for Predicting Patient-Specific
Hemodynamics of an Aortic Coarctation 110
    Idit Avrahami

Traditional CFD Boundary Conditions Applied to Blood Analog Flow
through a Patient-Specific Aortic Coarctation 118
    *Xiao Wang, D. Keith Walters, Greg W. Burgreen, and
    David S. Thompson*

Regular Papers

Extraction of Cardiac and Respiratory Motion Information
from Cardiac X-Ray Fluoroscopy Images Using Hierarchical Manifold
Learning 126
    *Maria Panayiotou, Andrew P. King, Kanwal K. Bhatia,
    R. James Housden, YingLiang Ma, C. Aldo Rinaldi, Jas Gill,
    Michael Cooklin, Mark O'Neill, and Kawal S. Rhode*

Dyadic Tensor-Based Interpolation of Tensor Orientation: Application
to Cardiac DT-MRI 135
    Jin Kyu Gahm and Daniel B. Ennis

Continuous Spatio-temporal Atlases of the Asymptomatic and Infarcted
Hearts 143
    *Pau Medrano-Gracia, Brett R. Cowan, David A. Bluemke,
    J. Paul Finn, Alan H. Kadish, Daniel C. Lee, João A.C. Lima,
    Avan Suinesiaputra, and Alistair A. Young*

Progress on Customization of Predictive MRI-Based Macroscopic
Models from Experimental Data 152
    *Mihaela Pop, Maxime Sermesant, Samuel Oduneye, Sudip Ghate,
    Labonny Biswas, Roey Flor, Susan Newbigging, Eugene Crystal,
    Nicholas Ayache, and Graham A. Wright*

Automatic Personalization of the Mitral Valve Biomechanical Model
Based on 4D Transesophageal Echocardiography   162
Jingjing Kanik, Tommaso Mansi, Ingmar Voigt, Puneet Sharma,
Razvan Ioan Ionasec, Dorin Comaniciu, and James Duncan

Fast Catheter Tracking in Echocardiographic Sequences for Cardiac
Catheterization Interventions   171
Xianliang Wu, R. James Housden, Niharika Varma, YingLiang Ma,
Kawal S. Rhode, and Daniel Rueckert

A Unified Statistical/Deterministic Deformable Model for LV
Segmentation in Cardiac MRI   180
Sharath Gopal and Demetri Terzopoulos

Multi-modal Pipeline for Comprehensive Validation of Mitral Valve
Geometry and Functional Computational Models   188
Dominik Neumann, Sasa Grbic, Tommaso Mansi, Ingmar Voigt,
Jean-Pierre Rabbah, Andrew W. Siefert, Neelakantan Saikrishnan,
Ajit P. Yoganathan, David D. Yuh, and Razvan Ioan Ionasec

Personalized Modeling of Cardiac Electrophysiology Using Shape-Based
Prediction of Fiber Orientation   196
Karim Lekadir, Ali Pashaei, Corné Hoogendoorn, Marco Pereanez,
Xènia Albà, and Alejandro F. Frangi

Automatic Extraction of the 3D Left Ventricular Diastolic Transmitral
Vortex Ring from 3D Whole-Heart Velocity-Encoded MRI Using
Laplace-Beltrami Signatures   204
Mohammed S.M. ElBaz, Boudewijn P.F. Lelieveldt,
Jos J.M. Westenberg, and Rob J. van der Geest

Direct Myocardial Strain Assessment from Frequency Estimation
in Tagging MRI.   212
Hanne B. Kause, Olena G. Filatova, Remco Duits,
L.C. Mark Bruurmijn, Andrea Fuster, Jos J.M. Westenberg,
Luc M.J. Florack, and Hans C. van Assen

Estimation of Electrical Pathways Finding Minimal Cost Paths
from Electro-Anatomical Mapping of the Left Ventricle   220
Rubén Cárdenes, Rafael Sebastian, David Soto-Iglesias,
David Andreu, Juan Fernández-Armenta, Bart Bijnens,
Antonio Berruezo, and Oscar Camara

Velocity-Based Cardiac Contractility Personalization
with Derivative-Free Optimization   228
Ken C.L. Wong, Maxime Sermesant, Jatin Relan,
Kawal S. Rhode, Matthew Ginks, C. Aldo Rinaldi,
Reza Razavi, Hervé Delingette, and Nicholas Ayache

Model-Based Estimation of 4D Relative Pressure Map from 4D Flow
MR Images 236
  Viorel Mihalef, Saikiran Rapaka, Mehmet Gulsun,
  Angelo Scorza, Puneet Sharma, Lucian Itu, Ali Kamen, Alex Barker,
  Michael Markl, and Dorin Comaniciu

Self Stabilization of Image Attributes for Left Ventricle Segmentation ... 244
  Sarada Prasad Dakua, Julien Abi-Nahed, and Abdulla Al-Ansari

A Framework for the Pre-clinical Validation of LBM-EP
for the Planning and Guidance of Ventricular Tachycardia Ablation 253
  Tommaso Mansi, Roy Beinart, Oliver Zettinig, Saikiran Rapaka,
  Bogdan Georgescu, Ali Kamen, Yoav Dori, M. Muz Zviman,
  Daniel A. Herzka, Henry R. Halperin, and Dorin Comaniciu

Image-Based Estimation of Myocardial Acceleration Using TDFFD:
A Phantom Study 262
  Ali Pashaei, Gemma Piella, Nicolas Duchateau, Luigi Gabrielli, and
  Oscar Camara

Author Index 271

Left Atrial Segmentation Challenge:
A Unified Benchmarking Framework

Catalina Tobon-Gomez[1], Jochen Peters[2], Juergen Weese[2], Karen Pinto[1],
Rashed Karim[1], Tobias Schaeffter[1], Reza Razavi[1,3], and Kawal S. Rhode[1]

[1] Division of Imaging Sciences & Biomedical Engineering, King's College London,
London, UK*
[2] Philips Technologie GmbH, Innovative Technologies, Forschungslaboratorien,
Hamburg, DE
[3] Department of Cardiology, Guys and St. Thomas NHS Foundation Trust,
London, UK*

Abstract. The knowledge of left atrial (LA) anatomy is important for
atrial fibrillation ablation guidance. More recently, LA anatomical models have been used for cardiac biophysical modelling. Segmentation of the
LA from Magnetic Resonance Imaging (MRI) and Computed Tomography (CT) images is a complex problem. We aimed at evaluating current
algorithms that address this problem by creating a unified benchmarking framework through the mechanism of a challenge, the *Left Atrial
Segmentation Challenge 2013* (LASC'13). Thirty MRI and thirty CT
datasets were provided to participants for segmentation. Ten data sets
for each modality were provided with expert manual segmentations for
algorithm training. The other 20 data sets per modality were used for
evaluation. The datasets were provided by King's College London and
Philips Technologie GmbH. Each participant segmented the LA including a short part of the LA appendage trunk plus the proximal parts of
the pulmonary veins. Details on the evaluation framework and the results obtained in this challenge are presented in this manuscript. The results showed that methodologies combining statistical models with region
growing approaches were the most appropriate to handle the proposed
task.

1 Introduction

Atrial fibrillation (AF) is the most common cardiac electrical disorder which
doubles the mortality rate of patients. It has been shown that ectopic beats
from within the pulmonary veins (PVs) commonly initiate AF [1]. Therefore,
catheter ablation strategies attempt to electrically isolate the PVs from the left
atrial (LA) body. Knowing the LA anatomy is crucial for the success of the

---

* This research was supported by the National Institute for Health Research (NIHR)
Biomedical Research Centre at Guy's and St Thomas' NHS Foundation Trust and
King's College London. The views expressed are those of the author(s) and not
necessarily those of the NHS, the NIHR or the Department of Health.

O. Camara et al. (Eds.): STACOM 2013, LNCS 8330, pp. 1–13, 2014.

CT DATASETS	MRI DATASETS

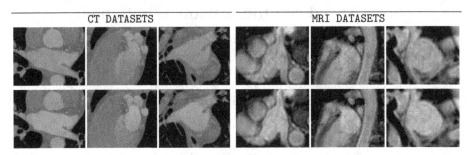

Fig. 1. Example datasets as provided for the challenge. Colour overlay shows the manual ground-truth, where green areas represent the LA body, and magenta areas represent the PVs (for details see Sec. 2).

intervention, since it enables accurate planning of ablation lines and guidance during the procedure [2]. More recently, LA anatomical models have been employed for cardiac biophysical modelling [3]. These models aim at understanding the mechanisms of AF and, eventually, at predicting optimal therapy.

Magnetic Resonance Imaging (MRI) and Computed Tomography (CT) are commonly used for imaging the heart. There are several topological variants of the LA and this means that segmentation of the endocardial boundary and PVs is a non-trivial task [4]. Therefore, we aimed at evaluating current algorithms that address this problem by creating a unified benchmarking framework through the mechanism of a challenge, the *Left Atrial Segmentation Challenge 2013* (LASC'13). This challenge was held at the MICCAI 2013 Workshop on Statistical Atlases and Computational Models of the Heart: Imaging and Modelling Challenges (STACOM'13). Each participant segmented the LA including a short part of the LA appendage trunk plus the proximal parts of the PVs, from 3D whole heart MRI and/or CT modalities. Details on the evaluation framework and the results obtained in this challenge are presented in this manuscript.

2   The Challenge

Thirty MRI and thirty CT datasets were provided to participants for segmentation. Ten data sets for each modality were provided with expert manual segmentations for algorithm training (see Fig. 1). The other 20 data sets per modality were used for evaluation. Datasets were limited to the most common topological variant showing four PVs. The datasets were provided by King's College London and Philips Technologie GmbH.

Participants were expected to segment the LA including a short part of the LA appendage (i.e. trunk) plus the proximal parts of the PVs (i.e. up to the first branching point or after 10 mm from the vein ostium). The LA body should have extended at least up to the mitral valve (MV) (i.e. reach into the funnel of the MV). Results were submitted as a single-valued binary mask covering all these structures in NifTI format[1].

---

[1] http://nifti.nimh.nih.gov/

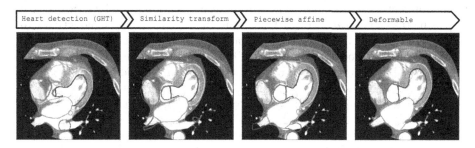

Fig. 2. Automatic segmentation pipeline on a CT image for ground truth generation. Different colours represent different parts of the deformable model. Green and magenta regions correspond to the LA and the PVs, respectively (for details see Sec. 3).

MRI Datasets. MRI acquisition was performed on a 1.5 T Achieva scanner (Philips Healthcare, The Netherlands). A 3D whole heart (3DWH) image was acquired using a 3D balanced steady-state free precession acquisition [5]. The sequence acquires a non-angulated volume covering the whole-heart with voxel resolution of $1.25 \times 1.25 \times 2.7\,\mathrm{mm}^3$. Images were acquired during free breathing with respiratory gating and at end-diastole with ECG gating. Main acquisition parameters include: TR/TE=4.4/2.4 ms, flip angle=90°, cardiac phases=1. Typical acquisition time for a complete volume is 10 min.

CT Datasets. Retrospectively ECG-gated cardiac multi-slice CT images were acquired with Philips 16-, 40-, 64- and 256-slice scanners (Brilliance CT and Brilliance iCT, Philips Healthcare, Cleveland OH, USA) at different cardiac phases. All images are reconstructed using a 512×512 matrix with an in-plane voxel resolution ranging from 0.30×0.30 to $0.78 \times 0.78\,\mathrm{mm}^2$ and with slice thickness ranging from 0.33 to 1.00 mm. All scans were acquired after injection of ca. 40–100 ml contrast media (density 320–370 mg iodine/ml), depending on the exact purpose of the study (assessment of coronary arteries or cardiac valves). Contrast levels vary widely over the images provided for this challenge. Acquisition times for a complete CT volume ranged from 3-5 sec on modern iCT scanners to 20 sec for the older 16-slice scanners.

3   Ground-Truth Generation

In order to obtain a set of ground-truth (GT) segmentations consistent across modalities, we started by performing an automatic model-based segmentation with a method which is optimised for both CT and MRI modalities. After the automatic segmentation, manual corrections were performed. Details are provided next.

Automatic Segmentation. The automatic segmentation used in this study was described in [7–9]. The segmentation uses shape-constrained deformable models. These are based on a mesh representation of surfaces of cardiac chambers and the attached great vessels. These meshes have a complex topology with T-junctions where different structures meet. The automatic adaptation starts by a localisation step using the Generalized Hough Transform [10] to place the mesh model close to the targeted organ. Thereafter, several adaptation steps with increasing degrees of freedom refine the model pose and shape. Each step uses trained boundary detectors that enable a robust and accurate detection of the wanted organ boundaries in the image. These detectors are trained individually per mesh triangle and can capture the varying appearance of organ boundaries in the images. Using the detected boundaries, a first step adjusts the global pose of the complete model by performing a rigid adaptation with scaling that minimises the squared distances of the model surface to the detected boundaries. Subsequent steps add more degrees of freedom by subdividing the model into mesh regions (such as cardiac chambers or short parts of the tubular vessels) and adapts these parts via individual affine transformations. Finally, a deformable adaptation step leads to a locally accurate segmentation where each mesh vertex is free to move under the image forces that pull the mesh triangles to the detected boundaries while internal forces regularise the adaptation and penalise strong deformations of the model shape. After adaptation of the model is complete, the regions enclosed by the surfaces are converted into a label image with region-specific labels. Labels not covering the LA and the PVs were discarded (see Fig. 2).

Manual Correction Criteria. Each automatic segmentation was manually corrected by an experienced observer to obtain the final GT segmentation. Manual corrections were performed using ITK-SNAP [11] for MRI datasets and Philips in-house editing tools for CT datasets. PVs were followed distally to the LA body ensuring at least 10 mm coverage. They were truncated at the branching point when there was no clear main PV to follow. This *early truncation* mainly happened in MRI, either due to image artefacts or low signal-to-noise ratio. Each obtained GT segmentation consists of five labels: one label for LA body and LA appendage trunk, and, four labels for each of the PVs. These labels were used for standardisation purposes (see Sec. 4).

4   Standardisation Framework

Even for a human observer, defining certain regions of the LA is difficult. One of these regions is the boundary between atrium and ventricle. Since the MV leaflets can be at different levels of *opening/closure*, the definition of a MV plane can be arbitrary. Unless an exact segmentation of the mitral annulus is available which, however, may be non-planar. Another one of these regions is the PVs. For these structures both the start points (i. e. ostia) and end points can be arbitrary. Finally, the LA appendage (LAA) varies greatly among the population which makes it difficult to segment. Since for most applications the actual shape of the LAA is not relevant, we opted for removing it from the LA body. We only retain the region most proximal to the LA cavity (i. e. trunk).

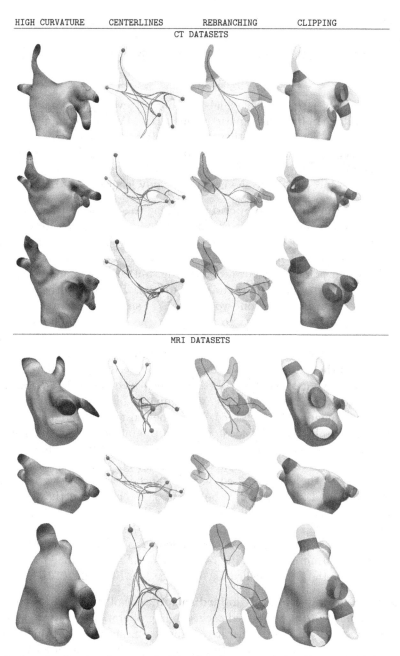

Fig. 3. Standardisation framework. Distal end points of the PVs and LAA were extracted by computing areas of maximum curvature. The centroids of the high-curvature areas were used as seed points for centerline extraction. The surface mesh was branched based on maximum inscribed sphere radius [6]. For PVs and LAA clipping: we computed a plane that is normal to the centerline and located 10 mm away from the ostia (for details see Sec. 4).

To ensure that the calculated metrics are not negatively affected by these regions, we standardised all submitted segmentations. The framework was implemented using the Visualization Toolkit (VTK), the Vascular Modeling Toolkit (VMTK), and MATLAB Toolbox Graph[2].

Mitral Valve. Given the predefined labels of our GT, we could compute certain anatomical landmarks in an automatic manner. For MV plane computation, we extracted a surface mesh representing the LA body and the PVs separately. We computed Principal Components Analysis on the LA body. The clip point was set along the main axis (or the average of the two main axes for more spherical bodies), at a distance of 35% × the maximum body length below the centroid.

Pulmonary Veins and LA Appendage. Along this clipped surface we computed the Gauss curvature [12,13]. We then normalised the curvature values to obtain a unified range of $(-2, 2)$. We thresholded the highest curvature values (>0.5). These patches of high curvature were used as candidate positions for PV and LAA end points (black contours in Fig. 3-HIGH CURVATURE). For each PV we selected the patch furthest from the ostium. The patches belonging to the body were discarded based on two criteria: *(1)* small surface area ($< 0.5 \times$ largest patch area); or *(2)* vicinity to the MV or the PVs ostia. The remaining body patches belonged to the LAA. From the selected patches we computed the centroids and stored them as *seed points* for centerline extraction.

We calculated the centerlines that connected each seed point to all remaining seed points plus the centroid of the MV edge, as displayed in Fig. 3-CENTERLINES. Using the approach of Antiga et al. [6], we computed bifurcation regions in the centerlines corresponding to each seed. From the most distal bifurcation point we defined a new splitting point located $0.75 \times$ the maximum inscribed sphere radius, similarly to the approach used by Piccinelli et al. [14] to define the neck of cerebral aneurysms (red section of centerlines in Fig. 3-REBRANCHING).

Next, we labeled the surface based on the branched centerlines. These automatically computed labels proved to be more consistent among the GT population than the arbitrarily defined manual labels. Therefore, we used them as a final definition of each anatomical region of the LA. Using the labels, we isolated each PV and clipped it with a plane perpendicular to its corresponding centerline and located 10 mm from the PV ostium. The LAA was clipped at 80% × the maximum length of the labelled LAA surface (see Fig. 3-CLIPPING).

Automatic Segmentations Standardisation. For all submitted segmentations (binary masks), we performed a close filling operation to ensure a single connected region. From it we generated a surface mesh using marching cubes followed by volume preserving smoothing. Next, we clipped the mesh with the MV plane generated from the GT mesh (Sec. 4), discarding unconnected regions. Then, we transferred the automatic branch labels of the GT mesh to their closest points in the automatically segmented mesh. For each label, we ensured a

---

[2] www.vtk.org www.vmtk.org www.ceremade.dauphine.fr/~
peyre/matlab/graph/content.html

single connected region to avoid transferring PV labels to neighbouring areas. Finally, using the labels we isolated each PV and LAA and clipped them using the planes computed automatically from the GT.

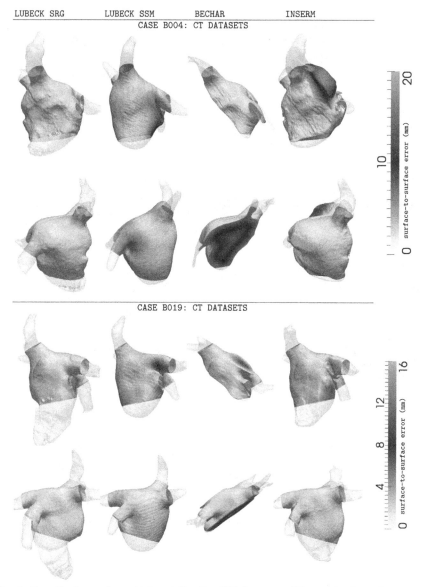

Fig. 4. Results for each participant for the CT datasets. The original meshes are displayed with transparency. The standardised meshes are colour mapped with surface-to-surface errors. Note that for visualisation purposes, only the automatic-to-GT errors are displayed in this figure. The symmetric surface-to-surface errors are summarised in Table 1 (for details see Sec. 5).

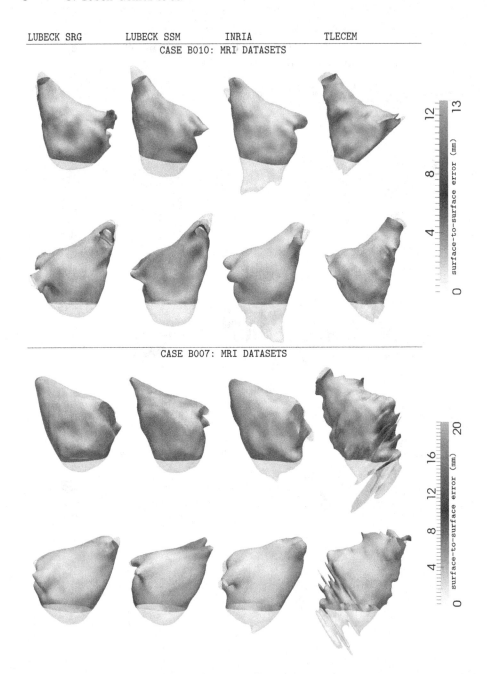

Fig. 5. Results for each participant for the MRI datasets. The original meshes are displayed with transparency. The standardised meshes are colour mapped with surface-to-surface errors. Note that for visualisation purposes, only the automatic-to-GT errors are displayed in this figure. The symmetric surface-to-surface errors are summarised in Table 1 (for details see Sec. 5).

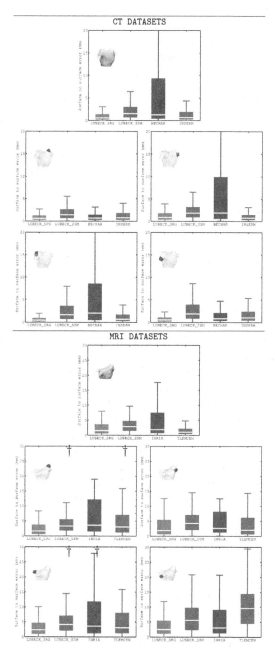

Fig. 6. Box-plots of segmentation errors for each participant for each anatomical region. The corresponding region is represented in the vignette on the upper left corner. Maximum whisker corresponds to approximately 99.3% coverage if the data were normally distributed. Pair of samples that yielded statistically non-significant differences ($p < 0.001$) are marked on the plot (†).

5   Discussion

Participants. We received submissions from 5 groups. University of Lubeck, Germany, processed CT and MRI datasets with two methodological approaches: one based on statistical shape models (LUBECK SSM), and, another one combining statistical shape models and region growing (LUBECK SRG). University of Bechar, Algeria, processed the CT datasets (19 out of 20) with a combination of region growing and gradient vector flow snakes (BECHAR). INSERM Rennes, France, processed the CT datasets with a multi-atlas, multi-voting and region growing approach (INSERM). Inria, Sophia-Antipolis, France, processed the MRI datasets using decision forests (INRIA). University of Tlemcen, Algeria, processed the MRI datasets with a combination of threshold localisation and circularity shape descriptors (TLEMCEN). For details on each methodology refer to the articles in these proceedings. Examples of the submitted segmentations before and after standardisation are shown in Fig. 4 and Fig. 5.

Results. To test segmentation accuracy, symmetric surface-to-surface error (S2S) and Dice metric were computed for all the standardised segmentations. The median and standard deviation of both metrics (i.e. LA body, left superior PV, left inferior PV, right superior PV and right inferior PV) are summarised in Table 1. Fig. 6 shows the box-plots of the S2S errors.

Results showed that statistical shape approaches combined with region growing obtained the best accuracy (LUBECK SRG and INSERM). It must be noted that this type of methodology often leaks into the left ventricle, the aorta and sometimes into the right atrium. Some of these segmentation errors were removed by our standardisation process, hence they were not penalised in by the evaluation metrics. However, to be implemented as a feasible clinical tool, the region growing should be somewhat constrained. For instance, limiting the region growing process to the PVs areas and/or to the surroundings of the initial surface. The statistical shape approach in itself (LUBECK SSM) although highly robust (i.e. valid shape instances of the LA) obtained lower accuracy than its corresponding extension with region growing. However, for certain applications this level accuracy could be sufficient.

BECHAR's approach shows potential since the slices that were processed obtained good accuracy. Unfortunately, due to the large amount of missing slices (specially on the lower part of the LA) the performance metrics were low in most anatomical regions. A 3D extension of the approach able to handle the whole span of the LA would increase the feasibility of the methodology. This was suggested by the authors as part of their future work.

INRIA's approach even though it makes few assumptions, often obtained good results. However, when then segmentation failed the errors where rather large resulting on poor average performance. A possible improvement of this approach would be to split the segmentation in two tasks: one for the body only and one for the PVs. Another improvement would be to impose shape constrains on the raw output of the decision forrest.

Table 1. Summary of error metrics: all structures measured after standardisation

		LUBECK SRG				LUBECK SSM				BECHAR				INSERM			
		s2s		dice		s2s		dice		s2s		dice		s2s		dice	
		m	std	m	std	m	std	m	std	m	std	m	std	m	std	m	std
CT	BODY	0.58	4.25	0.94	0.09	1.53	2.34	0.88	0.08	1.27	8.53	0.55	0.17	0.67	3.18	0.89	0.10
	LSPV	0.62	2.10	0.88	0.21	1.42	2.29	0.78	0.21	0.69	3.16	0.83	0.24	0.78	3.34	0.74	0.27
	LIPV	0.87	3.06	0.86	0.13	1.84	2.35	0.68	0.25	1.87	7.10	0.15	0.36	0.76	2.68	0.86	0.33
	RSPV	0.48	1.04	0.89	0.15	1.67	3.87	0.53	0.28	2.05	6.98	0.13	0.34	0.70	1.46	0.79	0.27
	RIPV	0.57	1.08	0.89	0.14	1.90	3.05	0.53	0.33	0.84	2.64	0.71	0.29	1.05	1.99	0.76	0.33
		LUBECK SRG				LUBECK SSM				INRIA				TLECEM			
		s2s		dice		s2s		dice		s2s		dice		s2s		dice	
		m	std	m	std	m	std	m	std	m	std	m	std	m	std	m	std
MRI	BODY	1.54	3.39	0.91	0.09	2.86	2.99	0.83	0.08	1.87	9.17	0.88	0.30	1.07	3.79	0.90	0.07
	LSPV	1.69	3.33	0.72	0.28	3.31	2.48	0.39	0.24	3.63	15.85	0.35	0.30	3.06	8.17	0.08	0.36
	LIPV	1.89	3.36	0.76	0.38	4.29	4.25	0.23	0.20	2.64	14.37	0.42	0.36	2.06	5.63	0.50	0.41
	RSPV	2.34	6.27	0.35	0.28	4.10	5.65	0.08	0.22	3.59	8.53	0.12	0.33	3.09	8.06	0.36	0.28
	RIPV	2.49	6.35	0.46	0.29	5.60	5.59	0.04	0.20	3.07	7.01	0.34	0.31	9.59	7.63	0.00	0.18

s2s = surface-to-surface error (mm); dice = Dice metric; m = median; std = standard deviation; LSPV = left superior PV; LIPV = left inferior PV; RSPV = right superior PV; RIPV = right inferior PV; BODY = LA body without LA appendage.

TLECEM's approach is based on circular shape descriptors from the sagittal plane. Thus it obtained good accuracy on the middle of the LA body (lowest error). However, the PVs were often missing and the lower part of the LA body (closer to the MV) was often over segmented. Similarly to the approach of BECHAR, a 3D extension could improve its feasibility for a clinical application.

6 Conclusions

This manuscript presents a unified benchmarking framework for current algorithms for segmentation of the left atrium from MRI and CT datasets. Strong effort was dedicated to implement a standardisation framework for the ground-truth and the automatic segmentations.

The results showed that methodologies combining statistical models with region growing approaches were the most appropriate to handle the proposed task. Visual results showed that an approach with good performance according to the error metrics (low surface-to-surface and high dice) does not always provide the best overall 3D structural result. This has pointed us to believe that it is important to explore other complementary metrics that better reflect the similarities in shape between the desired ground-truth surface and the automatic segmented surface.

As a follow-up work, we will submit a journal publication with the benchmarking framework presented in this workshop. In other to include other methodologies, we will make a second call for participants. In this follow-up work, we plan to evaluate the influence of image quality on each segmentation algorithm, include a measure of inter-observer variability, and, extend the performance metrics. Examples of these metrics include: centerline-to-centerline distance of the pulmonary veins, leakage metric to reflect the effect of a failed region growing, and, more advanced statistical measures of the shape differences between the surfaces.

Acknowledgements. The authors would like to thank C. Butakoff, O. Camara and A.J. Geers for their very useful suggestions for the automatisation of the evaluation framework.

References

1. Haïssaguerre, M., Jaïs, P., Shah, D.C., Takahashi, A., Hocini, M., Quiniou, G., Garrigue, S., Le Mouroux, A., Le Métayer, P., Clémenty, J.: Spontaneous initiation of atrial fibrillation by ectopic beats originating in the pulmonary veins. N. Engl. J. Med. 339(10), 659–666 (1998)
2. Calkins, H., Kuck, K.H., Cappato, R., Brugada, J., Camm, A.J., Chen, S.A., Crijns, H.J.G., Damiano, J.R.J., Davies, D.W., DiMarco, J., Edgerton, J., Ellenbogen, K., Ezekowitz, M.D., Haines, D.E., Haissaguerre, M., Hindricks, G., Iesaka, Y., Jackman, W., Jalife, J., Jais, P., Kalman, J., Keane, D., Kim, Y.H., Kirchhof, P., Klein, G., Kottkamp, H., Kumagai, K., Lindsay, B.D., Mansour, M., Marchlinski, F.E., McCarthy, P.M., Mont, J.L., Morady, F., Nademanee, K., Nakagawa, H., Natale, A., Nattel, S., Packer, D.L., Pappone, C., Prystowsky, E., Raviele, A., Reddy, V., Ruskin, J.N., Shemin, R.J., Tsao, H.M., Wilber, D.: 2012 hrs/ehra/ecas expert consensus statement on catheter and surgical ablation of atrial fibrillation: recommendations for patient selection, procedural techniques, patient management and follow-up, definitions, endpoints, and research trial design. Europace 14(4), 528–606 (2012)
3. Aslanidi, O.V., Colman, M.A., Stott, J., Dobrzynski, H., Boyett, M.R., Holden, A.V., Zhang, H.: 3d virtual human atria: A computational platform for studying clinical atrial fibrillation. Prog. Biophys. Mol. Biol. 107(1), 156–168 (2011)
4. Kato, R., Lickfett, L., Meininger, G., Dickfeld, T., Wu, R., Juang, G., Angkeow, P., LaCorte, J., Bluemke, D., Berger, R., Halperin, H.R., Calkins, H.: Pulmonary vein anatomy in patients undergoing catheter ablation of atrial fibrillation: lessons learned by use of magnetic resonance imaging. Circulation 107(15), 2004–2010 (2003)
5. Uribe, S., Muthurangu, V., Boubertakh, R., Schaeffter, T., Razavi, R., Hill, D.L.G., Hansen, M.S.: Whole-heart cine mri using real-time respiratory self-gating. Magn. Reson. Med. 57(3), 606–613 (2007)
6. Antiga, L., Steinman, D.A.: Robust and objective decomposition and mapping of bifurcating vessels. IEEE Trans. Med. Imaging 23(6), 704–713 (2004)
7. Peters, J., Ecabert, O., Meyer, C., Schramm, H., Kneser, R., Groth, A., Weese, J.: Automatic whole heart segmentation in static magnetic resonance image volumes. In: Ayache, N., Ourselin, S., Maeder, A. (eds.) MICCAI 2007, Part II. LNCS, vol. 4792, pp. 402–410. Springer, Heidelberg (2007)
8. Ecabert, O., Peters, J., Schramm, H., Lorenz, C., von Berg, J., Walker, M.J., Vembar, M., Olszewski, M.E., Subramanyan, K., Lavi, G., Weese, J.: Automatic model-based segmentation of the heart in CT images. IEEE Trans. Med. Imaging 27(9), 1189–1201 (2008)
9. Ecabert, O., Peters, J., Walker, M.J., Ivanc, T., Lorenz, C., von Berg, J., Lessick, J., Vembar, M., Weese, J.: Segmentation of the heart and great vessels in CT images using a model-based adaptation framework. Med. Image Anal. 15(6), 863–876 (2011)
10. Ballard, D.H.: Generalizing the Hough transform to detect arbitrary shapes. Pattern Recogn. 13(2), 111–122 (1981)

11. Yushkevich, P.A., Piven, J., Hazlett, H.C., Smith, R.G., Ho, S., Gee, J.C., Gerig, G.: User-guided 3d active contour segmentation of anatomical structures: significantly improved efficiency and reliability. Neuroimage 31(3), 1116–1128 (2006)
12. Alliez, P., Cohen-Steiner, D., Devillers, O., Lévy, B., Desbrun, M.: Anisotropic polygonal remeshing. ACM Transactions on Graphics 22, 485–493 (2003); SIGGRAPH 2003 Conference Proceedings
13. Cohen-Steiner, D., Morvan, J.M.: Restricted delaunay triangulations and normal cycle. In: 19th Annual Symposium on Computational Geometry, pp. 237–246 (2003)
14. Piccinelli, M., Veneziani, A., Steinman, D.A., Remuzzi, A., Antiga, L.: A framework for geometric analysis of vascular structures: application to cerebral aneurysms. IEEE Trans. Med. Imaging 28(8), 1141–1155 (2009)

Automatic Segmentation of the Left Atrium on CT Images

Abdelaziz Daoudi[1], Saïd Mahmoudi[2], and Mohammed Amine Chikh[3]

[1] University of Bechar
Road of Kenadsa, 08000 Bechar, Algeria
azizodaoudi@yahoo.fr
[2] Computer Science Department, Faculty of Engineering
University of Mons, 20, Mons 7000, Belgium
Said.Mahmoudi@umons.ac.be
[3] Biomedical Engineering Laboratory
Aboubaker Belkaid University
Tlemcen 13000, Algeria
mea.chikh@mail.univ-tlemcen.dz

Abstract. In this work, we present an automatic segmentation of the left atrium on computed tomography imaging (CT). The left atrium has an important role in patients with ventricular dysfunction as a booster pump to augment ventricular volume. A method based on active contours models with gradient vector flow is proposed in this paper and applied for left atrium segmentation. At first, a contrast enhancement is applied to improve the image quality. The automated initialization method is followed by a region-growing technique for a preliminary segmentation. The result of this technique is used as initialization for a segmentation method using a Gradient Vector Flow (GVF) snake based approach. The initial model can hence be attracted to the borders of the left atrium following various internal and external forces including the gradient vector flow (GVF).

Keywords: Left atrium, Adaptive histogram equalization, Region-growing, Snake, GVF, CT Cardiac images.

1 Introduction

Medical images segmentation consists of a set of methods used for extracting the relevant information in an automatic manner. This process improves the diagnoses made by several imaging techniques (CT, MRI, X-ray, etc.). The computed tomography imaging (CT) is one of the most popular imaging modality used to visualize the internal structures of the human heart. The left atrium is one of four chambers in the heart; it has an important role for patients with ventricular dysfunction as a booster pump to augment ventricular volume [1]. Recently, several studies have demonstrated how the left atrium plays a primary role not only in modulating ventricular filling and function through the atrioventricular interaction mechanism but also in providing important prognostic clues for the risk stratification of patients with diastolic dysfunction [2]. In the literature, many segmentation methods are used to detect the left atrium. In [3], the authors present a semi-automatic approach for segmenting the left atrium and the pulmonary veins from MR angiography (MRA). They use a region growing approach for locating the surrounding structures of the left atrium, and these

O. Camara et al. (Eds.): STACOM 2013, LNCS 8330, pp. 14–23, 2014.

regions are subdivided into disjoint regions based on their Euclidean distance transform, followed by a merge function which produces the segmented atrium. Some other methods in the literature are devoted to detect all of the pulmonary veins attached to the left atrium. For example, in [4] Depa et al. proposed to use a weighted voting label fusion to localize pulmonary veins. This work allows tracking center lines of the pulmonary veins entering the atrium. In [5], the authors used shape learning and shape-based image segmentation to identify the endocardial wall of the left atrium in the delayed-enhancement magnetic resonance images. An active contours approach was used to detect the endocardial and epicardial atrial wall segmentation on CT image in [6]. In [7], Koch et al. used the concept of Coherent Point Drift (CPD) registration for left atrium shape modeling, where a principle component analysis was applied in order to establish a deformable shape model.

In this paper, we propose a new segmentation method applied to the left atrium localization and delineation by using the Gradient Vector Flow (GVF) snake approach. In our proposal, we segment the left atrium with four pulmonary veins from Computed Tomography Imaging (CT) volume data sets. At first, an adaptive histogram equalization method was applied to improve the image quality. Then, a region-growing technique was used to initialize the GVF snake model. Finally, an automatic deformation of the GVF-snake model until the convergence was applied, allowing the final segmentation.

2   Proposed Method

The processing information frame work related to the proposed method is illustrated in figure (1). Our segmentation method is based on the GVF-snake model and allows an automatic detection of the left atrium outlines in CT image, figure (2). This detection method requires a good and accurate initialization of the model contour, ideally close to the area of the objects to be detected. After that, by using GVF internal and external forces, successively deformations are applied to the initial model curve by minimizing its related energy, so that it becomes closer to the real contour. In this method the initialization is done in an automatic way. The initialization is a result of the automatically localization of the left atrium in the image, followed by a region-growing segmentation technique. The dataset provided by STACOM 2013 workshop shows that the left atrium is in the middle of CT images, and it has a largest area. We will point in the following section that this characteristic is very important for the localization process.

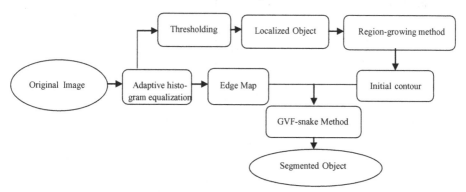

Fig. 1. The general framework of the proposed method

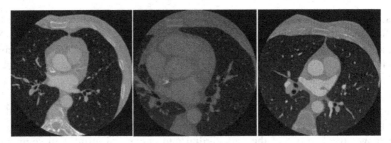

Fig. 2. Examples of CT images

2.1 Left Atrium Localization

Automatic initialization is a key to the automatic segmentation progress. So, in this section, we present a method based on an area metric to automatically select a point inside the left atrium. This method consists of the following steps:

- In the most cases, the left atrium is located in the middle of the CT image. The first of the localization method proposed is to reduce the image size to 260x260, (fig 4 a).
- After that, we apply at hresholding method to convert the initial image to a binary image. This thresholding permit to remove all objects that are smaller than a prede-fined threshold (fig 4 b). In this work, the thresholding is based on the gray level histogram analysis. The desired threshold should be the value that separates the left atrium from the background. The histogram image will determine a threshold or several thresholds depending on the nature of the image. In fact, if it is a bimodal image, the histogram allows to find a single threshold. If it is a multimodal image, we can find several levels; in this case we determine successively the best thre-shold. The threshold values correspond to the lowest part of the valley (fig 3). In this case the chosen threshold **T** is the one corresponding to the higher gray level value.
- At the final step, we compute the area of each object (fig 4c). The object with the largest area is recognized as the left atrium, and its centroid coordinate is used to initialize the region-growing technique (fig 4d).

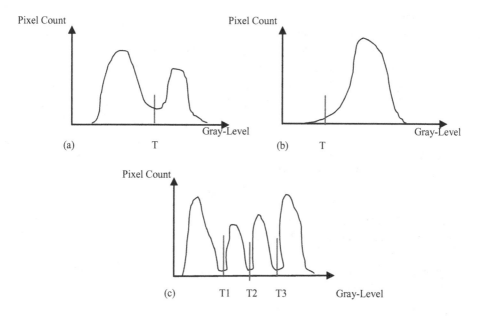

Fig. 3. Optimal threshold selection in gray-level histogram: (a) Bimodal, (b) Unimodal, (c) Multimodal

Fig. 4. LA location

(a)   Image reduced (260x260).
(b)   Binary image result.
(c)-(d)   LA centroid is labeled
       as a green symbol.

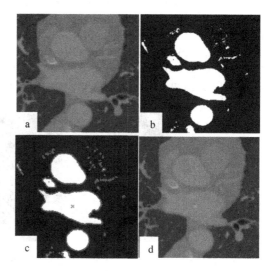

Once this seed point selected, the region growing process is started until it covers the entire region.

2.2   Region-Growing Method

The region-growing segmentation method was one of the first tools used for image segmentation, being a fast and intuitive technique. It is a simple approach for image segmentation and its principle is to gradually grow the region around its starting point (seed) [8]. It remains effective when its parameters are well chosen, as follow:

— The choice of seed.
— Fix a homogeneity criterium of the searched area, such as the difference between the gray level of the pixel and the average gray level of the region.
— Accumulation of neighboring pixels satisfying the homogeneity criterium.

Indeed, this method is very sensitive to the variation of these parameters and the final result is highly depending of them. Therefore, we need a rule describing a growth mechanism and a rule checking the homogeneity of the regions after each growth step.

In our work, in order to test the region homogeneity, the pixel intensity has to be close to the region mean value M (i), equation (1):

$$|I(x,y) - M(i)| <= T \qquad (1)$$

Where, T is the threshold predefined in the previous section.

On the other hand, the threshold T varies depending on the region and the intensity of the pixel I(x, y). The region growth until the distance between the region and possible new pixels become higher than the threshold T. The region growing algorithm that we used is presented as follows:

1. Start with a seed pixel.
2. Check the 8-neighbors of the pixel and analyze the homogeneity criterion.
3. Growth in the region until no pixel satisfies the criterion.

The goal of the region-growing technique is to find the initial contour that represents the meaningful model of left atrium. Therefore, the region-growing result was used as initial contour for the GVF snake model; since it provides an initial contour near to the left atrium borders. Figure (5) shows the original image on which was applied the region growing segmentation method.

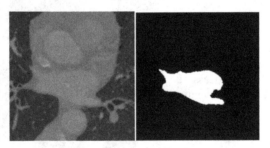

Fig. 5. Region-growing result

2.3   Left Atrium Segmentation

The major problems of CT images segmentation are resulting from the similarity between neighboring pixels of the organs in image. Therefore, it is necessary to go through a preprocessing to correct defects and improving the quality of the image before performing the segmentation. Adaptive histogram equalization (AHE) is one of

the well-known contrast enhancement techniques because it is simple and effective
[9]. The AHE can adjust the histogram so as to broaden the areas (edges) having a bad
distribution. Before the AHE, we subtract the threshold value of image; this difference is between the image background (background) and the interesting objects in the
image (foreground). The Adaptive histogram equalization is used to ameliorate the
image contrast, especially the boundaries of left atrium (fig 6).

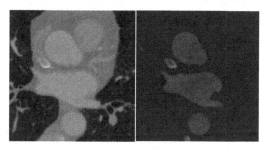

Fig. 6. Adaptive histogram equalization (AHE) result

Gradient Vector Flow

This section will describe in details the GVF-snake segmentation. Historically, traditional active contours were introduced by Kass, Witkin and Terzopoulos in 1988 [10].
They represent a method that minimizes the energy function consisting of an external
force (E_{ext}) and an internal force (E_{int}), equation (2):

$$E=\int_0^1 E_{int}(X(s)) + E_{ext}(X(s)) \tag{2}$$

Where s is the curvilinear abscissa along the contour $s \in [0, 1]$, and X denotes the
active contour. The internal energy (E_{int}) is defined by the following equation:

$$E_{int} = \frac{1}{2}(\alpha |X'(s)|^2 + \beta |X''(s)|^2) \tag{3}$$

Where α and β are the weighting parameters which control the snake's tension and
rigidity, and respectively, $X'(s)$ and $X'(s)$ denote the first and second derivatives of
$X(s)$ with respect to s. The external energy function (E_{ext}) is defined to be the negative
of image gradient magnitudes. The external energy E_{ext} is derived from the image so
that it takes on its smaller values at the features of interest, such as boundaries:

$$E_{ext} = -|\nabla I(x,y)|^2 \quad \text{or} \quad E_{ext}(s) = -|\nabla G_{\sigma *} I(x,y)|^2 \tag{4}$$

Where $I(x, y)$ represents the image, ∇ is the gradient operator, and G_σ is the
Gaussian filter with standard deviation σ. The total energy (E) should be:

$$E = \int_0^1 \frac{1}{2}[\alpha |X'(s)|^2 + \beta |X''(s)|^2] + E_{ext}(X(s))ds. \tag{5}$$

Xu and Prince [11] have proposed a new deformable model called Gradient Vector Flow snake. It was developed because of the known limitations of traditional snakes such as their poor convergence in concave regions. This method uses the GVF as a new external force by introducing a vector diffusion equation in order to diffuse the gradient of the edge map extracted from the image. This process starts by calculating the edge map of the given image. In our case, we have used a Sobel operator to this aim. The edge map characterizes the areas of interest in the image.

The GVF field is defined as a vector field $V(x, y) = (u(x, y), v(x, y)$ that minimizes the following energy function, equation (6).

$$\varepsilon = \iint \mu\,(ux^2 + uy^2 + vx^2 + vy^2) + |\nabla f|^2 |V - \nabla f|^2 dxdy. \tag{6}$$

Where f is the image edge map, μ is the smoothness degree, u and v represent the direction and strength of the field and ∇ f is the gradient of the edge map. The gradient ∇ f has vectors pointing towards the edge, which is a desirable property for snake.

Using the calculus of variations, the GVF field can be obtained by solving the Euler-Lagrange equations (7) and (8), and through the iteration of these, u, v will be obtained:

$$\mu\nabla^2 u - (u - fx)\,(f\,x^2 + f\,y^2) = 0 \tag{7}$$

$$\mu\nabla^2 v - (v - fy)\,(f\,x^2 + f\,y^2) = 0 \tag{8}$$

Where, ∇^2 is the Laplacian operator. The efficiency of snakes depends on a set of parameters such as μ (regularization parameter), alpha (elasticity parameter), beta (rigidity parameter), gamma (viscosity parameter), Kappa (external force weigh) and iteration number. And their values are shown in the experimental result, see Fig (9). The regularization parameter should be set according to the amount of noise present in the image (more noise, increase) (μ=0.1~0.2). The deformation of the GVF-snake is an iterative process. The iterations are stopped when a maximum number of iterations are reached.

3   Experimental Results

A set of 20 cardiac CT images provided by STACOM 2013 workshop have been tested by the proposed method in order to evaluate its performance. We implemented an algorithm of GVF snake in order to segment the left atrium of the heart in a CT image. Figure (7) shows how to locate the left atrium in the image. Two kinds of CT images are shows in fig (7a). Each image is represented by four images that are the original image reduced, the histogram image, the result of thresholding and the result of the localization. The histograms of the images are given in Fig (7b). The gray-level values corresponding to the valley of the histogram at (45.0, 78.0) and (76.0, 129.0, 152.0) are marked with the red points and we choose the best threshold of them. The binarization of the original image using the obtained threshold is given in fig (7c). The object with the largest area is recognized as the LA, and its centroid coordinate (x, y) is used as the seed pixel for following technique Fig (7d).

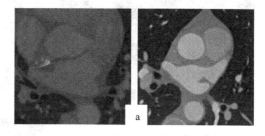

Fig. 7. LA location procedure

(a) Original image.
(b) Histogram and threshold values, (T_2, T3 are the best thresholds in red color).
(c) Binary image.
(d) LA centroid is labeled as a green point.

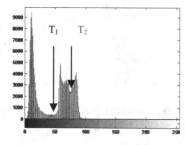
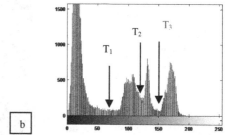

The initialization of the GVF snake is done by using of the region-growing technique, and the initialization is done closer to the left atrium boundary as shown in figure (8). Fig (8b) shows the region growing results. Fig (8c) shows the initial GVF snake in red color. From these experiments we can see that the initialization of the GVF snake by region growing algorithm plays a very important role in the segmentation process.

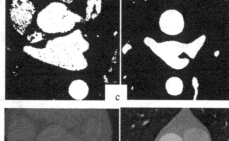
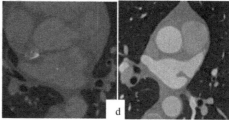

Fig. 8. GVF snake initialization

(a) Seed pixel in green color.
(b) Region growing results.
(c) Initial contour for the GVF snake model.

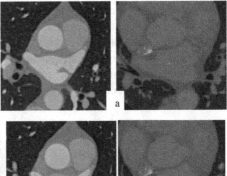

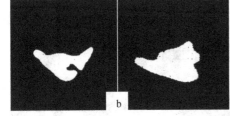

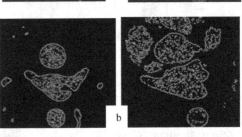

Fig. 9. Segmentation results

(a) AHE result.
(b) Edge map.
(c) Final contour snake in blue color.

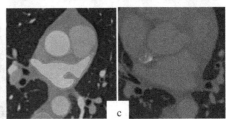

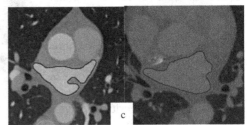

Figure (9), (a) shows the adaptive histogram equalization (AHE) result, the edge map is shown in (b) and the final snake in blue color after five iterations is shown in (c).

The value of alpha, beta, gamma and Kappa are respectively 1, 0.5, 1.4 and 1.2 for the GVF-snake algorithm.

4   Conclusion

In this paper, we have presented a technique for segmenting the left atrium from CT images by a GVF snake model. The Region Growing technique (RG) is proposed for extracting only the initial contour of the GVF snake. The threshold of RG is selected as a function of the histogram image. The proposed method is simple and fast. In some slices, the obtained results show that the results of segmentation include the detection of the pulmonary veins. We note that the segmentation result by GVF snake is better than the region growing technique taken as initialization model. The number of iterations required is greatly reduced through the use of the preliminary segmentation results where the average running time was 8 ±72 s. And the algorithm was implemented and executed in a 2.4 GHz Intel Core i3 PC with 4 GB RAM. In our future work, the proposed method will be extended to segment the left atrium in 3D image volume.

References

1. Stefanadis, C., Dernellis, J., Toutouzas, P.: A clinical appraisal of left atrial function. European Heart Journal 22, 22–36 (2001)
2. Sergio Macciò, M.D., Paolo Marino, M.: Role of the Left Atrium, pp. 53–70. Springer (2008)
3. Karim, R., Mohiaddin, R., Rueckert, D.: Left atrium segmentation for atrial fibrillation ablation. Proc. SPIE 6918, Medical Imaging (2008)
4. Depa, M., Sabuncu, M.R., Holmvang, G., Nezafat, R., Schmidt, E.J., Golland, P.: Robust atlas-based segmentation of highly variable anatomy: Left atrium segmentation. In: Camara, O., Pop, M., Rhode, K., Sermesant, M., Smith, N., Young, A. (eds.) STACOM 2010. LNCS, vol. 6364, pp. 85–94. Springer, Heidelberg (2010)
5. Gao, Y., Gholami, B., MacLeod, R.S., Blauer, J., Haddad, W.M., Tannenbauma, A.R.: Segmentation of the Endocardial Wall of the Left Atrium using Local Region-Based Active Contours and Statistical Shape Learning. Proc. of SPIE 7623, 76234Z-1 (2010)
6. Koppert, M.M.J., Rongen, P.M.J., Prokop, M., ter Haar Romeny, B.M., van Assen, H.C.: Cardiac left atrium CT image segmentation for ablation guidance. 978-1-4244-4126-6/10/$25.00 © IEEE (2010)
7. Koch, M., Bauer, S., Hornegger, J., Strobel, N.: Towards Deformable Shape Modeling of the Left Atrium Using Non-Rigid Coherent Point Drift Registration, pp. 332–337. Springer, Heidelberg (2013)
8. Adams, R., Bischof, L.: Seeded region growing. IEEE Trans. Pattern Anal. Machine Intell. 16(6), 641–647 (1994)
9. Wang, Z., Tao, J.: A Fast Implementation of Adaptive Histogram Equalization. 0-7803-9737-1. IEEE (2006)
10. Kass, M., Witkin, A., Terzopoulos, D.: Snakes: Active Contour Models. In: Proc. 1st International Conf. on Computer Vision, pp. 259–268 (1987)
11. Xu, C., Prince, J.L.: Snakes, Shapes, and Gradient Vector Flow. IEEE Transactions on Image Processing 7(3) (March 1998)
12. Zhang, M., Li, Q., Li, L., Bai, P.: An Improved Algorithm Based on the GVF-Snake for Effective Concavity Edge Detection. Journal of Software Engineering and Applications 6, 174–178 (2013)

Multi-atlas-Based Segmentation of the Left Atrium and Pulmonary Veins

Zulma Sandoval, Julián Betancur, and Jean-Louis Dillenseger

Inserm, U1099, Rennes, F-35000, France
Université de Rennes 1, LTSI, Rennes, F-35000, France

Abstract. This paper presents a multi-atlas segmentation approach concerning the left atrium and pulmonary veins in pre-operative CT images in order to plan ablation therapy in patients with atrial fibrillation. The segmentation procedure is composed of an atlas-based segmentation followed by a region-growing method. The atlas-based segmentation step exploits the a priori knowledge of existing structures to extract the inner region of the left atrium. The output of the atlas-based segmentation is then eroded to be used as seed volume in the region-growing procedure. This step adds new voxels according to a criterion which uses the intensity information of the input image.

Keywords: Multi-atlas segmentation, affine registration, elastic registration, region-growing.

1 Introduction

Atrial fibrillation is a cardiac arrhythmia caused by abnormal electrical discharges in the atrium. Ablation procedures have proved to be some of the most effective methods in treating of atrial fibrillation [1]. They aim to destroy mechanisms that trigger abnormal electrical charges or to modify the substrate that allows arrhythmia to be induced or maintained. Segmentation of the left atrium and pulmonary veins in pre-operative 3D-CT images is essential in planning correct ablation therapy.

Research interest in left atrium and pulmonary veins (LAPVs) segmentation has recently increased due to pulmonary veins appearing to play a key role in the initiation and maintenance of atrial fibrillation [2]. However, it is a challenging task due to the large anatomical variations, especially in the number, form and location of pulmonary veins. Some methods have been proposed for LAPVs segmentation. Some of them use prior knowledge of the anatomical shape such as model-based [3, 4] and atlas-based [5] approaches. Other approaches such as thresholding, region-growing and active contours use only the intensity information of the image [6, 7].

We propose an approach that combines prior knowledge of the anatomical shape to obtain the inner region of interest, with an intensity-based region-growing algorithm to extract particular anatomical details.

O. Camara et al. (Eds.): STACOM 2013, LNCS 8330, pp. 24–30, 2014.

2   Methods

The proposed method computes a detailed segmentation of the LAPVs from a coarse estimation of the atrium. This coarse segmentation uses a multi-atlas approach to obtain the inner region of the LAPVs. A region-growing based approach then performs a fine delineation from the inner region. Flowchart of the multi-atlas approach used is shown in (Fig. 1). This approach has already been used in the segmentation of the brain [8], prostate [9], whole heart and great vessels [5].

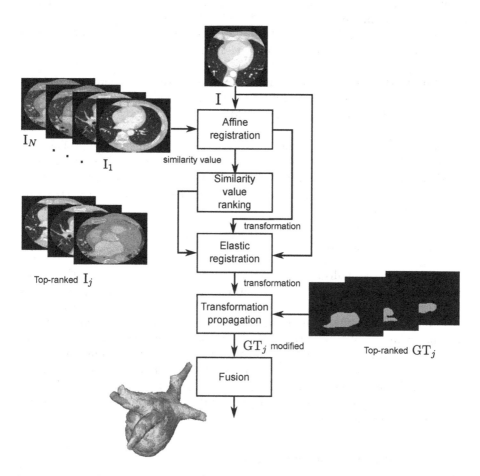

Fig. 1. Multi-Atlas based segmentation process. The affine registration allows to evaluate the similarity between the input image and the atlas data set. The elastic transformation modifies the ground truth of the selected atlas images. The fusion rule defines the coarse segmentation of the left atrium and pulmonary veins.

2.1 Multi-atlas Based Segmentation

Let I be an input CT image to be segmented and $A_1, A_2, ..., A_N$, a set of N atlases composed of intensity image I_i and the corresponding ground truth GT_i, as $A_i = (I_i, GT_i)$, $i = 1, 2, ..., N$. The coarse segmentation of the LAPVs is obtained with a multi-atlas approach composed of three steps: (i) atlas selection, (ii) elastic registration and (iii) fusion. The atlas selection step first performs an affine registration between I and each of the images composing the atlas dataset I_i. The images A_i are then ranked according to the value of the final similarity measure. Then, an elastic registration is performed between I and top-ranked I_i. The results of elastic registration propagated on the corresponding top-ranked GTs are fused to obtain the segmentation of I. The affine registration is automatically initialized. The translation needed to align the geometrical center of the fixed and moving images is used as an initial transformation.

Similarity Measures. Three similarity measures were considered for the registration process: sum of squared differences (SSD), normalized correlation coefficient (NCC) and mutual information (MI).

Fusion. A simple majority-voting fusion rule is used to merge the propagation of the labelled images of the selected atlas. This rule establishes that a voxel in I is labelled as a left atrium or pulmonary vein if at least half of the output elastic transformation of top-ranked labelled volumes are part of the structure at the voxel's location.

2.2 Region-Growing

The output of the multi-atlas segmentation is eroded to lie inside the region of interest. This output is used to initialize a region-growing approach. The seed region is shown in Fig. 2. This approach adds new voxels according to a criterion using the intensity information. In such a way, the seed region adapts to the particularities of each patient. An sphere structuring element with a radius of 2 mm was used to erode. Also, the growing procedure was constrained to a region corresponding to the dilated atlas-based segmentation output using the same structuring element. The intensity values of seed region are used to set automatically the criterion of growing. The interval is defined by two standard deviation around the mean value.

3 Experiments and Results

3.1 Data

The image dataset is composed of 30 CT images. Ten of them have a corresponding delineation of the left atrium and pulmonary veins (Ground Truth) and were used to train the atlas. The remaining 20 were used to evaluate performance.

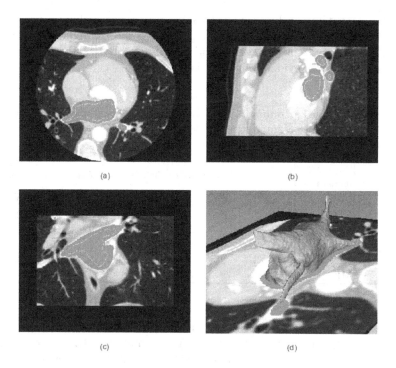

Fig. 2. The axial (a), sagittal (b) and coronal (c) views of the intensity input image (gray level), the corresponding ground truth (green) and the contour of the seed region (yellow). The model (d) of the seed region.

These images were provided by King's College London and Philips Research Hamburg to the participants of the Left Atrium Segmentation Challenge (LASC).

3.2   Evaluation Using the Training Images

To evaluate the performance, we used a leave-one-out approach to tune the atlas. The accuracy of both affine registration and elastic registration were assessed. Three similarity measures were tested: SSD, NCC and MI. The incidence of using a region-of-interest (ROI) containing the LAPVs was also evaluated. Then, two tests resulted: (i) using the ROI in both the input image and the atlas images (ii) without using ROI.

The accuracy of the registration was measured using the Dice similarity index. This index depicts the overlap of two volumes $V1$ and $V2$ and is defined as:

$$Dice = \frac{2|V_1 \cap V_2|}{|V_1| + |V_2|} \tag{1}$$

Affine Registration. Table 1 presents the results of the ranking process when A_1 is taken from the training group as input image and the remaining nine are used as atlas. It can be observed the incidence of using ROI. Indeed, the change in the ranking is more evident when MI is used.

Table 1. Images ranked according to the final value of the similarity measure in affine registration to A_1 (chosen as input image). Results for SSD, NCC and MI with/without ROI are shown.

ROI	Similarity measure	1st	2nd	3rd	4th	5th	6th	7th	8th	9th
with	SSD	2	3	7	6	5	4	8	9	10
	NCC	2	7	6	4	5	8	3	9	10
	MI	4	3	6	5	7	8	9	10	2
without	SSD	5	2	6	3	8	9	7	10	4
	NCC	5	3	2	6	9	7	8	4	10
	MI	5	10	9	6	3	8	7	4	2

Dice values between A_1 and the corresponding ranked images are presented in Table 2. In general, affine registration reaches an average Dice score between 33% and 82%. A low Dice index would lead to a weak initialization of next stage (elastic registration). This is the case of using MI and no ROI which obtained the worst Dice. It can be observed that using a ROI improves the performance.

Fusion. Table 3 contains the Dice indexes when two to seven images are used in the fusion. In general, Dice indexes are above 90% and are improved when using a ROI. The best performance was obtained for SSD. Despite the low affine Dice index for MI without ROI, the fusion rule increased the index for this case to a value comparable to the other metrics.

Table 2. Dice index between the ground truth of A_1 and the propagated atlas ground-truth using output affine transformations, for ranking images. Results for SSD, NCC and MI with/without ROI are shown.

ROI	Similarity measure	1st	2nd	3rd	4th	5th	6th	7th	8th	9th
with	SSD	82.0	80.3	64.1	79.5	72.6	74.7	67.1	77.5	66.2
	NCC	81.9	65.3	79.2	74.5	72.9	68.1	9.8	77.2	71.2
	MI	81.6	83.3	82.4	75.2	66.7	77.8	81.5	69.1	1.2
	Mean	81.8	76.3	75.2	76.4	70.7	73.5	52.8	74.6	46.2
without	SSD	72.0	67.1	75.4	26.3	63.4	57.7	68.0	37.9	59.6
	NCC	70.6	68.1	66.3	75.0	58.4	65.3	63.5	58.9	37.6
	MI	54.7	35.5	41.3	67.3	21.6	51.2	67.3	0.8	10.7
	Mean	65.8	56.9	61	56.2	47.8	58.1	66.3	32.5	36

Table 3. Dice index between the ground truth of A_1 and fused atlas ground truth after elastic registration. Results for SSD, NCC and MI with/without ROI are shown.

ROI	Similarity measure	Number of images used in fusion					
		2	3	4	5	6	7
with	SSD	93.9	88.2	91.6	90.9	94.4	91.9
	NCC	92.7	92.5	94.2	93.5	93.4	92.0
	MI	93.2	92.1	93.3	91.5	93.0	92.3
without	SSD	85.6	93.8	92.1	94.2	93.6	92.9
	NCC	89.1	92.7	93.8	88.6	89.1	88.0
	MI	91.4	84.8	88.6	86.7	89.4	86.9

Region-Growing. Table 4 contains the Dice indexes when the region growing is initialized with the eroded atlas-based segmentation. Region-growing approach improved the Dice score for the tests without ROI especially when less of five images are used in the fusion. Best Dice (95%) was obtained for SSD without ROI and using five images in fusion. Dice values are high no matter the number of images used in the fusion.

Table 4. Dice index between the ground truth of A_1 and region-growing output. Results for SSD, NCC and MI with/without ROI are shown.

ROI	Similarity measure	Number of images used in fusion					
		2	3	4	5	6	7
with	SSD	93.6	87.7	91.2	90.5	93.6	91.3
	NCC	92.7	92.0	93.4	92.9	92.7	91.3
	MI	92.8	91.2	92.3	90.6	92.0	91.4
without	SSD	87.0	94.2	94.8	95.0	94.6	93.3
	NCC	90.7	93.9	94.6	88.9	89.6	88.7
	MI	93.1	84.9	89.5	87.0	90.1	87.4

4   Conclusion

The segmentation of LAPVs was presented first using an atlas-based segmentation to compute an approximation of the structures and then a region growing procedure to obtain the anatomical details. Three similarity measures were tested with/without using a ROI in the registration procedure. The atlas-based segmentation reached a Dice score of 94% using the SSD metric with or without ROI. The region growing approach slightly improved the Dice when no ROI was used. This approach also allowed to obtain a high Dice index no matter the number of images used in the fusion.

References

[1] Cox, J.: Atrial fibrillation II: Rationale for surgical treatment. The Journal of Thoracic and Cardiovascular Surgery 126, 1693–1699 (2003)

[2] Schwartzman, D., Lacomis, J., Wigginton, W.: Characterization of left atrium and distal pulmonary vein morphology using multidimensional computed tomography. Journal of the American College of Cardiology 41(8), 1349–1357 (2003)

[3] Zheng, Y., Jhon, M., Boese, J., Comaniciu, D.: Precise segmentation of the left atrium in C-arm CT volumes with applications to atrial fibrillation ablation. In: 9th IEEE International Symposium on Biomedical Imaging (ISBI), pp. 1421–1424 (2012)

[4] Ecabert, O., Peters, J., Schramm, H., Lorenz, C., Von Berg, J., Walker, M.J., Vembar, M., Olszewski, M.E., Subramanyan, K., Lavi, G., Weese, J.: Automatic Model-Based Segmentation of the Heart in CT Images. IEEE Transactions Medical Imaging 27(9), 1189–1201 (2008)

[5] Zuluaga, M.A., Jorge Cardoso, M., Modat, M., Ourselin, S.: Multi-atlas propagation whole heart segmentation from MRI and CTA using a local normalised correlation coefficient criterion. In: Ourselin, S., Rueckert, D., Smith, N. (eds.) FIMH 2013. LNCS, vol. 7945, pp. 174–181. Springer, Heidelberg (2013)

[6] Karim, R., Mohiaddin, R., Rueckert, D.: Left atrium segmentation for atrial fibrillation ablation. In: Proc. of SPIE Medical imaging (2008)

[7] Zhu, L., Gao, Y., Yezzi, A., MacLeod, R., Cates, J., Tannenbaum, A.: Automatic segmentation of the left atrium from MRI images using salient feature and contour evolution. In: 2012 Annual International Conference of the IEEE Engineering in Medicine and Biology Society (EMBC), pp. 3211–3214 (2012)

[8] Leung, K., Barnes, J., Modat, M., Ridgway, G., Bartlett, J., Fox, N., Ourselin, S.: Automated brain extraction using multi-atlas propagation and segmentation (MAPS). In: 2011 IEEE International Symposium on Biomedical Imaging: From Nano to Macro, pp. 2053–2056 (2011)

[9] Acosta, O., Simon, A., Monge, F., Commandeur, F., Bassirou, C., Cazoulat, G., de Crevoisier, R., Haigron, P.: Evaluation of multi-atlas-based segmentation of CT scans in prostate cancer radiotherapy. In: 2011 IEEE International Symposium on Biomedical Imaging: From Nano to Macro, pp. 1966–1969 (2011)

Model-Based Segmentation of the Left Atrium in CT and MRI Scans

Birgit Stender[1], Oliver Blanck[2,3], Bo Wang[4], and Alexander Schlaefer[1]

[1] University of Lübeck, Institute for Robotics and Cognitive Systems,
Medical Robotics Group, Germany
`{stender,schlaefer}@rob.uni-luebeck.de`
`http://www.rob.uni-luebeck.de/`
[2] University Clinic Schleswig-Holstein,
Departement for Radiation Oncology, Lübeck, Germany
[3] CyberKnife Center Northern Germany, Güstrow, Germany
[4] Institute of Biomedical Analytical Technology and Instrumentation,
Xi'an Jiaotong University, Xi'an, China

Abstract. Ablation is a minimal invasive interventional method used in cardiac electrophysiology. It is one option for the treatment of patients suffering from paroxysmal or persistent atrial fibrillation through pulmonary vein isolation. During the intervention endocardial surface potentials from a tracked mapping catheter are recorded with respect to a static patient specific surface geometry. The purpose of the presented work is to compare two different automatic segmentation methods working on both CT and MRI volumes. Segmentation of the left atrium is challenging because the shape variability is high. The use of statistical shape models initialized by means of affine image registration was explored as first method. The second method was non-parametric and based on atlas registration and statistical region growing. Segmentation results were validated and compared using a leave-one-out cross validation on the volumes provided with segmentation results achieved manually by experts. The Dice's coefficient was used as error measure. The method based on statistical region growing performed better than statistical shape models. A Dice's coefficient of 0.87 was achieved on both imaging modalities.

1 Introduction

At present the most frequently performed electrophysiological cardiac intervention is ablation therapy for the treatment of atrial fibrillation (AFib). AFib seems to be often caused by abnormal sources of electrical excitation around one or several of the four pulmonary veins leading into the left atrium (LA) [5]. The resulting impairment of an organized atrial contraction is leading to an increased risk of thrombus formation in the left atrium and the left atrial appendage (LAA). In cases of paroxysmal or persistent atrial fibrillation one treatment option is isolation of the ectopic foci achieved by drawing circumferential ablation lines around the pulmonary veins (pulmonary vein isolation, PVI). The catheters approaching the atrial wall to be ablated are displayed in Fig. 1a). During PVIs Electroanatomical Mapping (EAM) systems like Carto™ (Biosense Webster, Diamond Bar, CA, USA) or EnSite NavX™ (St. Jude Medical, St. Paul,

O. Camara et al. (Eds.): STACOM 2013, LNCS 8330, pp. 31–41, 2014.
© Springer-Verlag Berlin Heidelberg 2014

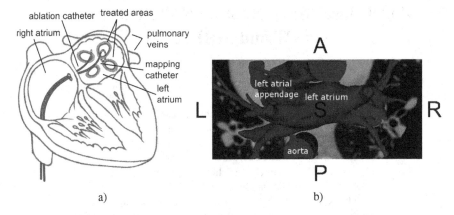

a)   b)

Fig. 1. a) Target region at the ostia of the pulmonary veins. Ablation and mapping catheter are entering the left atrium penetrating the atrial septum at the fossa ovale. b) Anatomical structures at risk illustrated by segmentation of the blood pool within CT dataset A001.

MN, USA) allow to navigate and record endocardial surface potentials with respect to a static patient specific surface geometry of the blood pool. Because EAM systems allow to navigate the tracked catheters almost in real time the X-ray exposure time caused by fluoroscopic guidance could be reduced for the time period after surface registration.

The accuracy requirements for segmentations of the left endocardial contour are challenging, first, because anatomic structures at risk are located close to the target region and second, because the ablation lines need to be drawn with high precision. As illustrated in Fig. 1b) anatomical structures at risk are the left atrial appendage (LAA) and the aorta. The esophagus not visible in Fig. 1b) is potentially also located close to the left atrial wall.

Improvement in left atrial segmentation could be achieved by addressing the following two sub-problems:

1. Fully automatic segmentation of the atrial endocardial contour in contrast-enhanced CT could reduce the preparation time and increase the reproducibility of this first step within the clinical workflow.
2. The substitution of CT with MRI as preoperative imaging modality would be desirable to further reduce the X-ray exposure of the patients. Currently the quality of the segmentation results is not as good as the one achieved in CT volumes. A further improvement of segmentation algorithms working in MRI and a detailed investigation of the results in direct comparison with CT data would allow an evidence-based decision making between the two imaging modalities.

Among the most recently published papers on left atrial segmentation the majority was focusing on MRI. Several authors pointed out that the contour of the left atrium is highly variable [4,12,8]. A relatively simple approach presented was based on segmentation of the blood pool and subsequent determination of cutting planes at narrowings [7]. Due to the inhomogeneous and patient-specific distribution of imaging gray values within the blood pool user-interaction was required. A segmentation algorithm based on registration of an atlas and labelmap fusion was already presented at the STACOM conference in 2010 by Depa et al. [4]. A very recently presented approach is based on splitting the complex structure of the contour into simpler substructures. The parts are then segmented using multi-model statistical shape knowledge [12,8].

We will present two different approaches. The first one is based on statistical shape models (SSM), the second one uses region growing. Both approaches utilize atlas information for initialization and work fully automatically.

2   Material and Methods

2.1   Material

The two segmentation approaches were tested with the 30 CT and 30 MRI volumetric scans of the Left Atrial Segmentation Challenge 2013 database. The imaging data was recorded using cardiac gating and contrast enhancement. Ground truth is provided through segmentations performed by clinical experts. The training datasets A for CT and MRI include 10 volumetric scans and ground truth labelmaps for each modality. The remaining 20 volumetric scans each given without ground truth will be referred in the following as CT dataset B and MRI dataset B. The mean spatial resolution is 0.45 mm for the CT and 1.25 mm for the MRI volumetric scans. Further information on the patients' anamnesis and the imaging protocols used was not provided.

2.2   Modality Specific Atlases of the Left Atrium

The quality of segmentation results achieved with SSM in general strongly depends on accurate initialization [6]. Usually the initial pose (position and orientation) of the mean shape is selected interactively which can be time-consuming and is potentially leading to user-dependent segmentation results. To overcome these known drawbacks we automatically determined the pose for the initial contour by means of atlas registration. This technique was used as well to initialize the statistical region growing approach. For this second approach seed points and assumptions on the range of gray values were automatically generated.

Two modality specific atlases were determined from CT dataset A and MRI dataset A. The volumes were cropped around the LA annotations within the corresponding labelmap. One of the LA volumes was selected as reference. It will be annotated as *ref* in the following. Pairwise affine registration was performed to map all LA volumes $i \neq ref$ in dataset A onto LA volume *ref*.

An affine registration algorithm is optimizing 12 parameters describing position, orientation, scaling and shearing by minimizing the mutual information similarity measure between two volumes. The result is a 4 x 4 transformation matrix $^{ref}H_i$ which describes the transformation of voxels from image volume i to image volume *ref*. We used the Fast Affine Registration and ResampleVolume2 modules released as part of 3D Slicer 3.6.3 for affine registration throughout the whole work [9]. Slicer modules were called via shell scripts. Histogram equalization and averaging of the transformed image volumes resulted in a mean volume used as template. These steps were performed in Matlab.

In Fig. 2 the result for CT is shown. The left (right) column images are all displayed with respect to the same coronal (transverse) cutting plane. The reference volume is displayed in Fig. 2 a) and b), the mean volume used as template in c) and d) and the probability density distribution in e) and f).

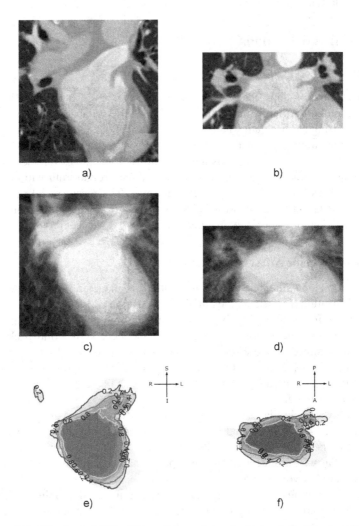

Fig. 2. a) and b): CT volume A003 used as reference dataset *ref*. c) and d): Atlas template volume created from the CT training dataset A. e) and f): Propability density distribution of the volume to belong to the blood pool of the left atrium.

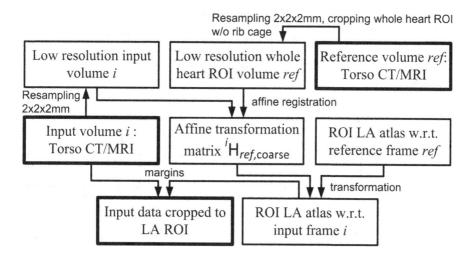

Fig. 3. Workflow for left atrial ROI detection in a CT or MRI input volume to be segmented

2.3   Identifying the Left Atrial Region of Interest

The workflow for detection of the LA ROI is displayed Fig. 3. In a first step the whole torso reference volume *ref* and the input torso volume i to be segmented were downsampled isotropically to a resolution of 2 mm. Within the downsampled reference volume an ROI including a large region of the heart is cropped without parts of the rib cage. The rib cage is excluded because the dimensions of the skeletal bones and the dimensions of the heart are not correlated in deseased patients.

Affine registration of the heart region to the input dataset resulted in the transformation matrix $^iH_{ref,coarse}$. This transformation matrix describes the position, orientation, scaling and shearing of the reference image volume *ref* with respect to the input image volume i. Because the modality specific atlas was determined with respect to the reference volume the margins of the atlas volume were also defined with respect to this image volume. The rectangular LA atlas ROI was transformed with the affine transformation matrix $^iH_{ref,coarse}$ to determine the LA ROI with respect to the volume i. Within the input volume in its original resolution the LA ROI was then cropped as starting point for the initialization of the two segmentation algorithms.

2.4   Statistical Shape Models

SSM are a set of methods which has already been successfully applied to various medical image segmentation tasks. While active shape models (ASM) include prior knowledge of the mean shape and its variations, active appearance models (AAM) are additionally augmented with statistical knowledge of imaging features associated with the contour [3,2,6].

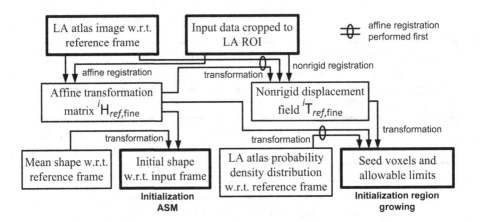

Fig. 4. Initialization workflow for both segmentation algorithms. By means of atlas registration the initial contour for the segmentation approach using a statistical shape model was determined as well as the initial volume and margins for segmenation based on statistical region growing.

Initialization Based on Image Registration. The segmentation algorithm started with an affine registration of the atlas image onto the input volume as shown in Fig. 4. The mean contour was afterwards mapped using the same affine transformation matrix. This initial contour guess is constrained by the permitted variations of the SSM.

The Point Distribution Model. For representation of the mean shape and shape variability a point distribution model (PDM) was used. Such a PDM is built up in the following steps: The groundtruth annotations within all N image volumes are described by a certain number k of landmark points on the contour. The $3k$ land-mark point coordinates for each contour i are listed in the column vector $\mathbf{x}_i = (x_1, \ldots, x_k, y_1, \ldots, y_k, z_1, \ldots, z_k)$. Assuming known point-to-point correspondences among the landmark points the mean shape is then determined by simply averaging over the $3k$ point coordinates:

$$\bar{\mathbf{x}} = \frac{1}{N} \sum_{i=1}^{N} \mathbf{x}_i \tag{1}$$

By means of an eigendecomposition on the sample covariance matrix

$$\mathbf{S} = \frac{1}{N-1} \sum_{i=1}^{N} (\mathbf{x}_i - \bar{\mathbf{x}}) (\mathbf{x}_i - \bar{\mathbf{x}})^T, \tag{2}$$

the eigenvectors ϕ_m and the corresponding eigenvalues λ_m are determined. Permiss-able shape variations are modeled as linear combination of the eigenvectors with the largest eigenvalues.

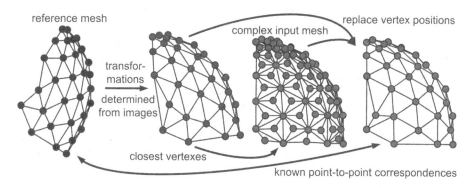

reference mesh   complex input mesh   replace vertex positions

transfor-
mations

determined
from images

closest vertexes

known point-to-point correspondences

Fig. 5. Surface transformation based on image registration

Point-to-Point Correspondences. Setting up the PDM presupposes known point-to-point correspondences between the different surface meshes. The shapes further need to be aligned which is usually done with Procrustes or generalized Procrustes analysis (GPA) based on the vertex points. We used instead the affine transformation matrices determined during atlas construction as illustrated in Fig. 5. From the ground truth labelmap of the reference dataset (A003 for CT, A002 for MRI) a simplified surface mesh was created (2000 vertexes for CT, 500 vertexes for MRI). Within Fig. 5 this mesh is represented by the one on the left hand side with red vertexes. From the other groundtruth labelmaps complexer surface meshes (10,000 vertexes for CT, 2500 vertexes for MRI) were computed. One of these meshes is the one with green vertexes in Fig. 5. The same affine registration algorithm and settings as used for building the atlases were applied to register the reference volume onto each of the nine other volumes included in dataset A. Subsequently non-rigid transformation fields were computed with the same demons registration algorithm and settings as used for registering the atlases onto a previously unseen volume. The resulting affine transformation matrix and nonrigid transformation field were used for warping the vertexes of the simplified reference mesh. The operation is displayed in Fig. 5 as transformation from the red mesh to the blue one. For reasons of clarity the blue mesh is not displayed superimposed with the green mesh. For each vertex included in the blue mesh the closest vertex within the green mesh was identified (green vertexes with red borders). The vertex coordinates of each blue vertex were than replaced with the ones of the closest vertex. The connectivity remained unchanged.

Shape Adaptation. After initialization the contour is propagated in each adaptation step along the surface normals of the vertexes. The step size is determined based on gradient features. After adaptation the shape is tested for successful representation within the shape space confidence intervals.

2.5   Stastical Region Growing

Statistical region growing is a segmentation algorithm based on a search algorithm. For all voxels within an intermediate segmentation result the mean intensity value is determined. The voxels within a certain neighborhood of the intermediate segmentation result are listed in a queue. A voxel from the queue is inserted into the intermediate segmentation result if its intensity value difference from the mean is within certain limits. Its voxel neighborhood is then added to the queue. The algorithm terminates if the queue is empty. Instead of using a single voxel as seed point we started with all voxels having a high probability to be part of the LA.

The probability distribution was determined by atlas registration using affine and susequent non-rigid registration as illustrated in Fig. 4. The Fast Symmetric Forces variant of Thirion's demons algorithm was used as non-rigid registration algorithm [10,11]. The basic algorithm uses the following modified version of the optical flow equation for the pixel shift $\mathbf{u}$

$$\mathbf{u} = \frac{(m - f)\nabla f}{|\nabla f|^2 + (m - f)^2},$$ (3)

where m and f denote the intensity values of the fixed and moving image. The term $(m - f)^2$ has a stabilizing effect. For the affine and subsequent non-rigid registration we called the Fast Affine Registration and BRAINSDemonWarp module released as part of 3D Slicer 3.6.3 via shell scripts [9].

Instead of updating the limits within the statistical region growing algorithm at each iteration the mean and standard deviation of the image gray values were determined from the set of seed voxels. A 26-voxel neighborhood was used. The threshold around the mean image gray value was set to $\pm 1.5\sigma$ for CT and $\pm 3\sigma$ for MRI. We implemented the algorithm in C++ and called it as Matlab mex-file. Afterwards the filter module Voting Binary Hole Filling Filter within 3D Slicer 3.6.3 was applied [9]. The parameters selected were: majority threshold 1, maximum radius 6 for CT and 3 for MRI.

3   Results

3.1   Cross Validation

For quantitative comparison of the segmentation results we used the Dice's coefficient $dc\,(\mathbf{SR}, \mathbf{GT})$ as similarity measure between the set of segmented voxels $\mathbf{SR}$ and the set of ground truth voxels $\mathbf{GT}$. To explore the capability of initialization based on image registration we first validated results from affine and non-rigid registration without any further segmentation steps. Each ground truth labelmap was therefore transformed to the grid of the remaining images within training dataset A. The transformations applied resulted from registration of the underlying images. The Dice's coefficients from registration are listed in line (a) and (b) of Tab. 1. The mean for affine registration was 0.64 for CT ad 0.74 for MRI. These values were increased by 8.8% for MRI and by 22.1% for CT trough additionally applied non-rigid registration. The standard deviation was also raised from 0.10 (CT) and 0.06 (MRI) to 0.11 for both imaging modalities.

Using statistical region growing with atlas based initialization in a leave-on-out cross validation scheme the mean Dice's coefficient could further be increased to 0.87 for both imaging modalities in comparison to affine and subsequent non-rigid registration only. For further details please refer to Tab. 1 line (c). The standard deviation of the dice coefficient was reduced to a third. The segmentation results achieved with SSM

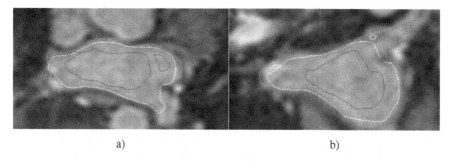

a)   b)

Fig. 6. Segmentation result achieved with statistical region growing on MRI volume A001 (dice coefficient 0.88) displayed for an axial cutting plane (a) and a coronal cutting plane (b). Red contour: Seed voxels as given by the registered probability density distribution of the atlas with value equal to 1. White contour: Segmentation result. Green contour: Ground truth provided.

(Tab. 1 line (d)) were again similar for both imaging modalities and a did not outperform the results from statistical region growing (0.83 mean Dice's coefficient for MRI and CT).

3.2   Computational Complexity

The computing times required are listed in Tab. 2. All computations were performed on a Windows XP Professional x64 operating system. The CPU was an Intel Xeon CPU 5160 running at 3.00 GHz. 15.9 GB of RAM was installed. For statistical region growing initialization by means of atlas registration (Step: Determination of seed voxels) accounted for 82.2% (CT) and 96.8% (MRI) of the computational effort.

Table 1. Dice's coefficients for the left atrium in CT and MRI using only registration (affine (a), affine and subsequent non-rigid (b)) in comparison to results achieved with the two segmentation approaches ((c) and (d))

		mean	std	min	max
(a) Affine registration	CT	0.6383	0.0965	0.1846	0.8266
	MRI	0.7414	0.0639	0.5598	0.8447
(b) Non-rigid registration	CT	0.7794	0.1053	0.4619	0.9373
	MRI	0.8070	0.1062	0.0412	0.9121
(c) Statistical region growing	CT	0.8674	0.0289	0.8123	0.9038
	MRI	0.8706	0.0333	0.8084	0.9001
(d) Statistical shape model	CT	0.8271	0.0528	0.7307	0.8808
	MRI	0.8303	0.0559	0.7003	0.9024

Table 2. Computing time for the different steps of the segmentation algorithms given in sec.

		CT	MRI
Statistical region growing	Determination of seed voxels	383.96	15.06
	Region growing	0.98	0.03
	Hole Filling Filter	82.00	0.47
	Sum	**466.94**	**15.56**
Statistical shape models	Affine transformation	16.39	6.57
	Model adaptation	215.52	13.55
	Affine Backtransformation	16.39	6.57
	Sum	**248.30**	**26.69**

This was mainly caused by non-rigid image registration of the atlas volume using the original imaging resolution (up to 0.3 mm for CT, up to 0.75 mm for MRI). For SSM the largest share of the computational costs was caused by the iterative shape model adaption step (86.8% for CT, 50.8% for MRI). In total the computational time for SSM was 46.8% lower for CT and 71.5% higher for MRI in comparison to statistical region growing.

4   Discussion and Conclusion

We demonstrated two methods for automatic segmentation of the endocardial contour of the left atrium. The approaches are applicable to CT and MRI. The first method is based on image registration and statistical region growing. A seed voxel volume and a mask for the image volume containing the left atrium were generated based on atlas registration. This approach is non-parametric because it is not based on a certain parameterization of the contour.

The second approach is using SSM. The shape and its variations are explicitly represented by means of a point distribution model. The property of being non-parametric is an advantage in case of high variability of the anatomical shape. As already pointed out at the STACOM 2010 by Depa et al. this is especially the case for the anatomy of the pulmonary veins leading into the left atrium [4]. Because two veins can built a common trunk not even the number must necessarily be constant. In addition comparison of ground truth contours achieved within MRI and CT images revealed that in CT the contours typically propagated further into the pulmonary veins but to different extents within the volumes included in dataset A.

The number of volumes given with corresponding ground truth is quite limited and therefore most likely not a sufficient representation of the whole space of shape variations. This is reflected by the fact that statistical region growing outperformed the approach based on SSM with respect the Dice's coefficient. However, there are also limitations of statistical region growing, as indicated by leaking within the left ventricle or the aortic root in some image volumes. Furthermore a higher computing time was required for initialization in CT images.

Within the clinical workflow the computational complexity of the algorithms is not the most important feature. Typically CT or MRI images are recorded on the day before. In addition preparation of the patients in the catheter lab takes several minutes.

The computing time of the non-rigid registration included might be reduced by downsampling the CT volumes in this step without substantially affecting the quality of the segmentation results afterwards. The results also indicate, that it may be interesting to study a hybrid approach including statistical region growing around the pulmonary veins after convergence of the shape adaptation of the SSM. This would allow to segment a larger range of anatomical variations among pulmonary vein anatomy while still preventing leaking into the aortic root and left ventricle by the restrictions modeled within the shape space.

Acknowledgements. This work was partially supported by BMBF grant 01EZ1140A.

References

1. Aliot, E., Ruskin, J.N.: Controversies in ablation of atrial fibrillation. European Heart Journal Supplements 10(supp. H), H32–H54 (2008)
2. Cootes, T.F., Edwards, G.J., Taylor, C.J.: Active appearance models. Pattern Anal. Mach. Intell. 23(6), 681–685 (2001)
3. Cootes, T.F., Taylor, C.J., Cooper, D.H., Graham, J.: Active shape models - their training and application. Comput. Vis. Image Und. 61(1), 38–59 (1995)
4. Depa, M., Sabuncu, M.R., Holmvang, G., Nezafat, R., Schmidt, E.J., Golland, P.: Robust Atlas-Based Segmentation of Highly Variable Anatomy: Left Atrium Segmentation. In: Camara, O., Pop, M., Rhode, K., Sermesant, M., Smith, N., Young, A. (eds.) STACOM 2010. LNCS, vol. 6364, pp. 85–94. Springer, Heidelberg (2010)
5. Hassaguerre, M., Jas, P., Shah, D.C., Takahashi, A., et al.: Spontaneous initiation of atrial fibrillation by ectopic beats originating in the pulmonary veins. N. Engl. J. Med. 339(10), 659–666 (1998)
6. Heimann, T., Meinzer, H.-P.: Statistical shape models for 3D medical image segmentation: A review. Medical Image Analysis 13(4), 543–563 (2009)
7. John, M., Rahn, N.: Automatic left atrium segmentation by cutting the blood pool at narrowings. In: Duncan, J.S., Gerig, G. (eds.) MICCAI 2005. LNCS, vol. 3750, pp. 798–805. Springer, Heidelberg (2005)
8. Kutra, D., Saalbach, A., Lehmann, H., Groth, A., Dries, S.P.M., Krueger, M.W., Dössel, O., Weese, J.: Automatic Multi-model-Based Segmentation of the Left Atrium in Cardiac MRI Scans. In: Ayache, N., Delingette, H., Golland, P., Mori, K. (eds.) MICCAI 2012, Part II. LNCS, vol. 7511, pp. 1–8. Springer, Heidelberg (2012)
9. Pieper, S., Lorensen, B., Schroeder, W., Kikinis, R.: The NA-MIC Kit: ITK, VTK, Pipelines, Grids and 3D Slicer as an Open Platform for the Medical Image Computing Community. In: Proceedings of the 3rd IEEE International Symposium on Biomedical Imaging: From Nano to Macro, vol. 1, pp. 698–701 (2006)
10. Prima, S., Thirion, J.-P., Subsol, G., Roberts, N.: Automatic analysis of normal brain dissymmetry of males and females in MR images. In: Wells, W.M., Colchester, A.C.F., Delp, S.L. (eds.) MICCAI 1998. LNCS, vol. 1496, pp. 770–779. Springer, Heidelberg (1998)
11. Vercauteren, T., Pennec, X., Perchant, A., Ayache, N.: Diffeomorphic Demons: Efficient Non-parametric Image Registration. Neuroimage, 45(1 suppl.), S61–S72 (2009)
12. Zheng, Y., Wang, T., John, M., Zhou, S.K., Boese, J., Comaniciu, D.: Multi-part left atrium modeling and segmentation in C-arm CT volumes for atrial fibrillation ablation. In: Fichtinger, G., Martel, A., Peters, T. (eds.) MICCAI 2011, Part III. LNCS, vol. 6893, pp. 487–495. Springer, Heidelberg (2011)

Toward an Automatic Left Atrium Localization Based on Shape Descriptors and Prior Knowledge

Mohammed Ammar[1,*], Saïd Mahmoudi[2], Mohammed Amine Chikh[1], and Amine Abbou[3]

[1] Biomedical Engineering Laboratory, University of Tlemcen, Algeria
ammar.mohammed4@gmail.com,
mea_chikh@mail.univ-tlemcen.dz
[2] University of Mons, Faculty of Engineering, Computer Science Department. 20 Place du parc,
Mons, B-7000, Belgium
Said.Mahmoudi@umons.ac.be
[3] Department of Cardiology, Tlemcen University Hospital, Tlemcen, Algeria
abbou.amine@yahoo.fr

Abstract. The left atrium is one of the four chambers of the heart. It receives oxygenated blood from the lungs and pumps it into the left ventricle. This blood is then circulated to the rest of the body. In a healthy adult the left atrium pumps blood into the ventricle in a regular rhythm. In atrial fibrillation (AF), the left atrium quivers in an abnormal rhythm and is no longer able to pump blood into the left ventricle efficiently. On the other hand MRI and CT are commonly used for imaging this structure. Segmentation can be used to generate anatomical models that can be employed in guided treatment and also more recently for cardiac biophysical modelling. For this reason, segmentation of the left atrium is a task with important diagnostic power. In this paper, we propose an automatic localization method in order to detect the left atrium in MRI images. Our method is based on shape descriptor and prior knowledge. For this purpose some descriptors are selected: circularity, area, the center of mass of each region, elongation factor, type factor. We propose also to use some prior knowledge as pulmonary artery position, and the left atrium position.

Keywords: Left atrium, shape descriptors, prior knowledge, localization.

1 Introduction

Cardiovascular diseases are the most common causes of deaths in the word. Heart strokes and attacks are two pathologies that affect the left atrium. The American Heart Association reports that 15% of all heart strokes are caused by a life threatening condition called atrial fibrillation (AF) [1].

Recently, a number of segmentation algorithms have been developed to detect LA in MRI or CT images.

For example in[2]the authors present a semi-automatic approach for left atrium segmentation and the pulmonary veins from MR angiography (MRA) data sets. They also propose an automatic approach for further subdividing the segmented atrium into

O. Camara et al. (Eds.): STACOM 2013, LNCS 8330, pp. 42–48, 2014.
© Springer-Verlag Berlin Heidelberg 2014

the atrium body and the pulmonary veins. The idea of this segmentation algorithm is that in MRA images the atrium becomes connected to surrounding structures via partial volume affected voxels and narrow vessels, thus the atrium can be separated if these regions are characterized and identified. The blood pool, obtained by subtracting the pre- and post-contrast scans, is segmented using a region growing approach and subdivided into disjoint subdivisions on the basis of the of the Euclidean distance transform. These subdivisions are then merged automatically starting from a seed point and stopping at the points where the atrium leaks into a neighbouring structure. The resulting merged subdivisions produce the segmented atrium. As second technique they propose an automatic approach used to identify the atrium body from segmented left atrium images. The separating surface between the atrium body and the pulmonary veins gives the ostia locations and can play an important role in measuring their diameters.

Another automatic approach for LA segmentation on cardiac magnetic resonance images was presented in [3]. This method used a weighted voting label fusion and a variant of the demons registration algorithm adapted to handle images with different intensity distributions to segment LA. In another paper[4], Yefeng et al. have proposed a segmentation approach applied toun-gated C-arm CT, where thin boundaries between the LA blood pool and surrounding tissues are often blurred due to the cardiac motion artifacts. The segmentation of this kind of images presents a big challenge compared to the highly contrasted gated CT/MRI. To avoid segmentation leakage, the shape prior was exploited in a model based approach to segment LA parts. However, independent detection of each part was not optimal and its robustness needs further improvement (especially for the appendage and PVs). So, they proposed to enforce a statistical shape constraint during the estimation of pose parameters (position, orientation, and size) of different parts. In[5]the authors present a method used to extract heart structures from CT and MRA data sets, in particular the left atrium. First, the segmented blood pool was subdivided at narrowings in small components. Second, these basic components were merged automatically so that they represent the different heart structures. The resulting cutting surfaces have a relatively small diameter compared to the diameter of the neighboring heart chambers. Both steps are controlled by only one fixed parameter. The method was presented as being fast and allowing interactive post-processing by the user.

Other authors in [6] proposed to use shape learning and shape-based image segmentation to identify the endocardial wall of the left atrium in the delayed-enhancement magnetic resonance images.

Some other works in the literature exploit a prior shape of the LA (either in the form of an atlas [7,8] or a mean shape mesh [9]) to guide the segmentation process. For example, Manzkeet al. [9] built a mean shape of the combined structure of the LA chamber and PVs from a training set. With a prior shape constraint, they could avoid the leakage around weak or missing boundaries, which plagues the non-model based approaches.

In this paper, we present a new approach to detect left atrium in MRI images. Our method is based on shape descriptors and prior knowledge.

2   Methodology

2.1   Left Atrium Localization

This section focuses on left atrium localization. To achieve this goal we propose the following algorithm:

-   Threshold the original image: we start with a preliminary thresholding operation. The threshold value for each case is set empirically.
-   Choose a slice where left atrium and aorta have circular shape.
-   Process the binary image obtained: remove small regions, fill holes, separate objects.
-   Characterization with shape descriptors: circularity, elongation, area, center of masse X, center of masse Y.

3   Characterization of the Left Atrium

A characterization step is essential to identify the regions of interest in cardiac MRI images. In the field of pattern recognition, we can find a large number of descriptors. The choice of an appropriate one depends on the object to be characterized. For some slices, the left atrium and the pulmonary artery have a generally circular shape and the left atrium is under the pulmonary artery, for this reason we chose the following attributes:

Perimeter of the Region P (R): This descriptor is calculated as the sum of the distances between successive contour pixels.
A (R): Area of the region.

Heywood Circularity Factor: $FcH = P(R)/(2\sqrt{\pi. A(R)}$ (1)

Rectangularity: Is defined by the value R calculated by the following formula,
R= Object area/ area of the minimum rectangle supervision (RME),

Elongation Factor: Is defined by the following formula,= **EF= RME length / width of RME**

Type Factor:= Is a complex factor that relates the area to the moment of inertia.

$$F_t = \frac{A^2}{4\pi\sqrt{I_{xx}+I_{yy}}}$$ (2)

-   M_x Center of masse X $\frac{(\sum x)}{A}$
-   M_y Center of masse Y $\frac{(\sum y)}{A}$
-   Moment of inertia: $I_{xx} = (\sum x^2) - A * M_x{}^2$
-   Moment of inertia: $I_{yy} = (\sum y^2) - A * M_y{}^2$

4   Results and Discussions

To evaluate the influence of these parameters, we selected 20 images (10 from training data and 10 from test data). For each image, we applied morphological operators like border rejection and elimination of smalls regions.

On the other hand, we can see clearly that the left atrium is great than the pulmonary artery, and it is always at the bottom. So the center of the left atrium is at the bottom of the center of the Pulmonary artery. After the binarization, we estimate the center of masse X, center of masse Y, the factor type, Heywood Circularity factor, elongation factor and area. Above

Table 1 and 2 show the min, max values and the standard deviation for the parameters defined above for 20 images selected (10 from training and 10 from test)

Table 1. The min, max value and standard deviation for the LA

	Center of masse X	Center of masse Y	Area	Elongation factor	Heywood Circularity	Type factor
min	161.25	140.17	378	1.70	1.04	0.93
max	213.29	197.83	3054	2.72	1.33	0.98
STD	15.27	16.64	623.27	0.28	0.07	0.02

Table 2. The min, max value and standard deviation for the PA

	Center of masse X	Center of masse Y	Area	Elongation factor	Heywood Circularity	Type factor
min	149.39	111.71	183	1.55	1.00	0.88
max	214.38	155.47	865	2.33	1.17	0.99
STD	15.67	11.34	149.84	0.22	0.07	0.03

We notice from table 1 and 2 that the left atrium has an area greater than the PA. On the other hand, the PA has a circular shape. Indeed, the Heywood Circularity factor values are between the values 1 and 1.28, and the factor type value are generally greater than 0.93. On the basis of these results and using the position of the LA and PA we have proposed our algorithm presented in section 2 described above. We illustrate in figure 1 some results for patients: A002 slice 87, patient A001 slice 51 and patient A004 slice 83.

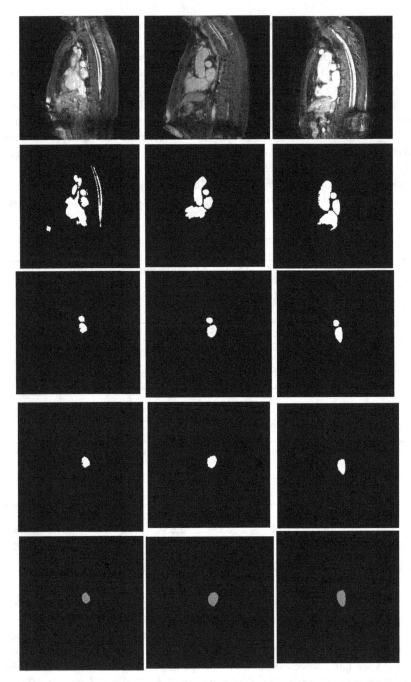

Fig. 1. Examples of segmentation and localization of LA using the proposed method: first line 3 MRI images. Second line : binarization. Third line: processing of the binary images in order to keep only LA and PA based on circularity index.   Fourth line : LA detected. Fifth line: Ground truth segmentations.

We present also in figure 2 an example from test data: B002 slice 93

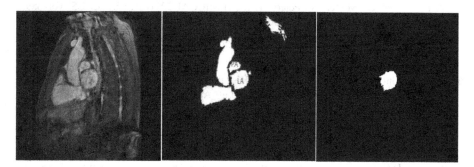

Fig. 2. Example of segmentation and localization of LA using the proposed method

For the 3D segmentation we use the center of mass of the LA detected to define a ROI around this center in order to localize the LA atrium in the next and previous slices.

5   Conclusion

We have presented in this paper a simple method used to localize the left atrium. The proposed algorithm was based on a preliminary threshloding and some morphological operations. In order to detect the left atrium we used some shape descriptors as position, area and circularity.

Our method was applied on the slices where the pulmonary artery has a circular shape. Once the PA artery is detected we localize the left atrium using the center of mass of X and Y.

For the others slices, we propose to use the center of gravity detected on the selected slice to search for the LA atrium region in previous and next slices.

At the moment the propose algorithms are not fully automatic because the threshold method provides sometimes bad results. That is why; we propose to combine in a future work the proposed detection method with a segmentation approach more efficient.

References

1. Atrial Fibrillation Investigators, Risk factors for stroke and efficacy of antithrombotic therapy in atrial fibrillation: analysis of pooled data from five randomized controlled trials. Archives of Internal Medicine 154(13), 1449-1457 (1994)
2. Karim, R., Mohiaddin, R., Rueckert, D.: Left atrium segmentation for atrial fibrillation ablation. In: Proc. of SPIE Medical Imaging (2008)
3. Depa, M., Sabuncu, M.R., Holmvang, G., Nezafat, R., Schmidt, E.J., Golland, P.: Robust atlas-based segmentation of highly variable anatomy: Left atrium segmentation. In: Camara, O., Pop, M., Rhode, K., Sermesant, M., Smith, N., Young, A. (eds.) STACOM 2010. LNCS, vol. 6364, pp. 85–94. Springer, Heidelberg (2010)

4. Zheng, Y., Wang, T., John, M., Zhou, S.K., Boese, J., Comaniciu, D.: Multi-part Left Atrium Modeling and Segmentation in C-Arm CT Volumes for Atrial Fibrillation Ablation. In: Fichtinger, G., Martel, A., Peters, T. (eds.) MICCAI 2011, Part III. LNCS, vol. 6893, pp. 487–495. Springer, Heidelberg (2011)

5. John, M., Rahn, N.: Automatic left atrium segmentation by cutting the blood pool at narrowings. In: Duncan, J.S., Gerig, G. (eds.) MICCAI 2005. LNCS, vol. 3750, pp. 798–805. Springer, Heidelberg (2005)

6. Gao, Y., Gholami, B., MacLeod, R.S., Blauer, J., Haddad, W.M., Tannenbaum, A.R.: Segmentation of the Endocardial Wall of the Left Atrium using Local Region-Based Active contours and Statistical Shape Learning. In: Dawant, B.M., Haynor, D.R. (eds.) Medical Imaging 2010: Image Processing. Proc. of SPIE, vol. 7623, 76234Z · © 2010 SPIE · CCC code: 1605-7422/10/$18 · doi: 10.1117/12.844321

7. Karim, R., Juli, C., Malcolme-Lawes, L., Wyn-Davies, D., Kanagaratnam, P., Peters, N., Rueckert, D.: Automatic segmentation of left atrial geometry from contrast-enhanced magnetic resonance images using a probabilistic atlas. In: Camara, O., Pop, M., Rhode, K., Sermesant, M., Smith, N., Young, A. (eds.) STACOM 2010. LNCS, vol. 6364, pp. 134–143. Springer, Heidelberg (2010)

8. Depa, M., Sabuncu, M.R., Holmvang, G., Nezafat, R., Schmidt, E.J., Golland, P.: Robust atlas-based segmentation of highly variable anatomy: Left atrium segmentation. In: Camara, O., Pop, M., Rhode, K., Sermesant, M., Smith, N., Young, A. (eds.) STACOM 2010. LNCS, vol. 6364, pp. 85–94. Springer, Heidelberg (2010)

9. Manzke, R., Meyer, C., Ecabert, O., Peters, J., Noordhoek, N.J., Thiagalingam, A., Reddy, V.Y., Chan, R.C., Weese, J.: Automatic segmentation of rotational Xray images for anatomic intra-procedural surface generation in atrial fibrillation ablation procedures. IEEE Trans. Medical Imaging 29(2), 260–272 (2010)

Decision Forests for Segmentation of the Left Atrium from 3D MRI

Ján Margeta[1], Kristin McLeod[1], Antonio Criminisi[2], and Nicholas Ayache[1]

[1] Asclepios Research Project, INRIA Sophia-Antipolis, France
[2] Machine Learning and Perception Group, Microsoft Research Cambridge, UK

Abstract. In this paper we present a method for fully automatic left atrium segmentation from 3D cardiac magnetic resonance datasets. We propose a machine learning approach using decision forests that requires very few assumptions on the segmentation problem. First, we extract the blood pool using a simple thresholding technique. Then, we learn to separate the left atrium from other structures in the image by using context-rich features applied on images enhanced with a multi-scale vesselness filter and transformed to measure distance to blood pool surface. We present our results on the STACOM LA Segmentation Challenge 2013 validation datasets.

1 Introduction

The left atrium plays an important role in facilitating uninterrupted circulation of oxygenated blood from the pulmonary veins to the left ventricle and in cardiac electro-physiology. To quantify its function, simulate electrical wave propagation and determine the best location for ablation therapy, it is important to be able to first accurately segment the atrial contours. A common approach to segment the left atrium from 3D images is to use statistical shape models [1][2]. A levelset based method with an heuristic region split and merge strategy was proposed by Karim et al. [3]. Finally, label fusion techniques [4] seem to yield accurate segmentations but require nonrigidly registering the image to be segmented to every training image. All these methods are specifically handcrafted for atrial segmentation and thus require treating the training set in a particular way or need a set of nonrigid registrations which can be computationally expensive.

The problem can be also formulated as a binary classification between atrial and background voxels. Similar to our previous work [5] and the work of Lempitsky et al. [6] for segmentation of left ventricles from dynamic magnetic resonance (MR) sequences and 3D ultrasound respectively, or segmentation of multiple sclerosis lesions [7] from multichannel MR, we propose a fully automated voxel-wise left atrium segmentation method based on classification forests [8]. The advantage of these approaches is that very few assumptions are necessary and it is possible to learn to segment directly from the image - label map pairs. The segmentation then consists of just a set of simple binary decisions independent of voxel position that can therefore be run rapidly in parallel for each voxel.

O. Camara et al. (Eds.): STACOM 2013, LNCS 8330, pp. 49–56, 2014.
© Springer-Verlag Berlin Heidelberg 2014

For this method we do not have the need for a robust registration method, to build a statistical model, nor explicitly define the classification problem and enforce the modality. We only require several training images with the atrium carefully delineated and good blood pool contrast in the images.

2 Dataset

The STACOM 2013 LA segmentation challenge dataset [9] contains 30 CT and 30 MR images. Each of these sets was divided into a training set (10 3D volumes with left atrial segmentation maps) and a validation set (20 3D volumes with no delineation provided). In this paper we used only the MR data. Compared to our previous work [5] there is no need for pose standardisation. The only preprocessing step was to linearly rescale the intensities between 0 and 98.5 percentiles. This was chosen to cut off noisy high intensity variation similarly to [10].

3 Decision Forests for Left Atria Segmentation

Decision forests are an ensemble supervised learning method consisting of a set of binary decision trees each composed from a set of simple binary decision functions. Our method consists of two phases. First, we train the structure of the decision forests using all available training images. This forest is then used to segment previously unseen datasets.

3.1 Training

In the offline training phase the structure of each tree is greedily optimised such that atrial voxels are split from the background voxels. This means in practice that at each node of the tree a randomly chosen subset of features is extracted for each voxel belonging to the node. Then for each feature θ a threshold τ maximising the information gain is selected among a regularly sampled range of meaningful thresholds. The best feature - threshold pair is then stored together as a binary splitting criterion ($\tau < \theta$) for the node and the voxels are split into the left and right disjoint partitions accordingly. This random feature sampling leads to increased inter-tree variability and better generalization [8]. The data division recursively continues until the chosen maximum depth is reached, the number of voxels reaching the node is too small or when a significant part of the voxels at the node already belongs to a single class and no significant information gain can be obtained by further splitting. These terminal nodes then become leaves of the tree. For each leaf, class distributions of voxels reaching the leaf can be easily obtained.

Training of random forests is relatively fast as each tree can be learnt individually in parallel. To further reduce the training time, to better balance the background/atrium voxel proportion and to improve discrimination of voxels on the boundaries, we train only on some of the voxels from the annotated set. As positive atrial voxel examples, all voxels in the training set annotated as atrium are taken. However, we sample the negative examples only on a sparse regular grid and add all voxels in the immediate atrial neighbourhood (approximately 15 pixels thick obtained by morphologically dilating the mask).

3.2 Segmentation Phase

During the segmentation phase each voxel is passed through the forest to reach a leaf in each tree. The average class distributions of all reached leaves then represents the posterior probabilities of the voxel belonging to either the atrium or the background given its appearance in the feature space. This means we obtain an atrium probability map for the whole volume. To obtain binary masks required for evaluation in the challenge, we simply need to threshold the atrial probability maps. Afterwards, we perform simple morphological hole filling on these thresholded images and extract the largest connected component to serve as the final binary segmentation.

3.3 Feature Families

To describe the appearance of each voxel and discriminate between the atrium and background we generate a random feature pool operating on the 3D images from two feature families as in Geremia et al. [7], applied on three different image channels (see section 3.4). The two feature families differ in how much local or remote information they capture around the tested voxel.

Local Features. Measure average channel intensity in the vicinity of the tested voxel (a $3 \times 3 \times 3$ cube centred on the voxel).

Context Rich Features. Defined as the difference between the tested voxel's channel intensity $I(x)$ and two remote region box channel intensity averages:

$$\theta_I(x) = I(x) - 0.5 \left(\frac{1}{Vol(R_1)} \sum_{x' \in R_1} I(x') + \frac{1}{Vol(R_2)} \sum_{x' \in R_2} I(x') \right) \tag{1}$$

The 3D regions R_1 and R_2 are randomly sampled in a relatively large neighbourhood around the tested voxel x. These features capture strong contrast changes but also long-range intensity relationships.

3.4 Image Channels

MR Image Intensity. Voxel intensity in combination with context rich features can give a wealth of information about its position. For example regions of the bright atrium are located close to the darker spine or lungs, and next to other

a) image + gt   b) blood pool   c) distance   d) tubularity

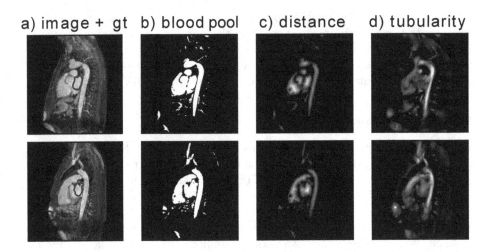

Fig. 1. Image channels extracted from two example images. a) Source image overlaid with the groundtruth segmentation. b) Blood segmented with sequentially applied Otsu thresholding. c) Distance to the blood contours. d) Tubularity enhancing vascular structures (note e.g. the strong signal in the aorta).

bright cavities. It is however much more difficult to discriminate between voxels on the border between the atrium and the ventricle as there is no clear intensity change (apart from occasional faint signal from the mitral valve). Therefore we extract the local and context rich features not only on image intensity (Fig. 1a), but also add two other channels (see Fig. 1c and Fig. 1d).

Distance to Blood Pool Contours. The atrium is always contained within the bright blood pool in the image. Thanks to the high contrast between the blood and the rest of the image, all blood can be extracted rather simply by sequentially applying Otsu's thresholding [11]. The first round splits the image between the air and the brighter part of the thorax. The second round then splits the brighter part of the thorax into the very bright blood and the rest (Fig. 1b).

We can observe that the atrium is mostly separated at blood pool narrowings (such as at the mitral plane or atrial septum). These can be located by measuring the distance to blood pool contours as distance minima (Fig. 1c). Local maxima are on the other hand located near centres of blobs in the image such as the atrium. Therefore, similarly to Karim et al. [3], we exploit these properties and compute the euclidean distance to the blood pool surface for each voxel in the image (voxels out of the blood pool are assigned zero distance). Instead of manually defining region splitting and merging criteria we let the forest pick the most discriminative decisions from the above mentioned feature families.

Tubularity Features. To further help in distinguishing the atrium from the other bright structures such as the aorta we calculate vesselness information for each voxel. This adds context based on enhanced arteries present (e.g. the atrium is near the aorta). We use a multiscale vesselness filter[12] enhancing all tubular objects ranging from 5 to 15 millimetres in diameter (see Fig. 1d).

4   Results

We trained a classification forest with previously described features. We chose the best parameters by running cross-validation on the training set. As the size of the training set is quite limited we used a leave-one-out approach i.e. we trained on 9 images and tested on the remaining one (Fig. 2). The best found settings were then applied to the validation data (Fig 3):

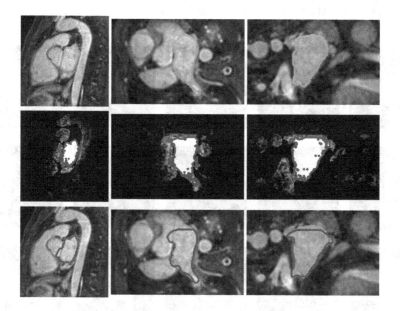

Fig. 2. Coronal, sagital and axial views of the segmentation results on one of the test cases during the leave one out cross validation. Top row: source image with ground truth, middle row: atrial probability map with contour of the probability map at 0.6 (brighter values means more confidence in the segmentation), bottom: source image with overlaid groundtruth (green) and final segmentation after hole filling (red).

Number of trees: 5 More trees result in increased accuracy and smoother segmentations but also increase training and classification time.

Number of features tested at each node: 200 Too few tested features results in more randomness in the forest but also less efficient splits, on the other hand higher numbers decrease generalization strength as the trees look more alike.

Number of thresholds tested for each feature value: 20 Similar to previous.

Maximal depth of the tree: 20 Deeper splits can better capture the structure, but can lead to overfitting.

Minimal number of points: 8 Too few of this parameter would result in noisy segmentations as even a single training voxel can significantly influence the result.

Neighbourhood in which context rich features are sampled: 70x70x70 Larger neighborhoods can capture more context, but result in frequently clipped context rich features as they get evaluated out of image bounds.

Binarizing threshold: 0.9 This parameter is used to keep only the more confident voxels and reduce the effect of oversegmentation.

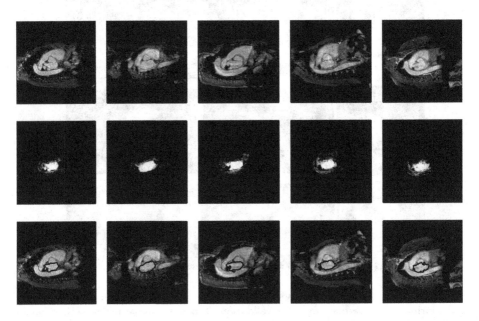

Fig. 3. Qualitative display of segmentation results on a subset of the validation dataset shown on mid-atrial slices. Top row: source images, middle: corresponding atrial probability maps, bottom: overlaid final segmentations.

After the selection of parameters via cross-validation we obtained a dice coefficient 0.632 ± 0.137. This small forest of 5 trees took on average 2 hours to train and just around a minute to fully automatically segment a single atrial image on a 12 core Intel Xeon 3.3GHz CPU.

This algorithm performs reasonably well to extract the main part of the atrial volume. The main drawbacks of our method are that the segmentation does not necessarily adapt to the cavity contours (see Fig. 3) and pulmonary veins are often missed or misclassified (see Fig. 4).

Fig. 4. Qualitative display of segmented atrial meshes from the validation dataset

5 Conclusions

We used a generic machine learning based image segmentation method and extended it with a set of image features to better distinguish vascular structures from the rest. We then learnt to directly predict voxel labels (atrium / background) from the images without hand-tuning the segmentation pipeline. The current solution is reasonably fast. Our current preliminary results could serve as an excellent atrium detector and initialization for a refinement step prior to the use in an accurate segmentation for electro-physiological studies.

Although in this paper we apply this method only to MR it is straightforward to apply to other modalities such as images from the computed tomography. We only assume strong blood pool contrast.

Due to the fact that the training set consists of only 10 images it does not cover many of the atrial shape variations present in the validation images. To capture some of this variability and avoid image registration to an atlas in the classification phase we aim to enrich the training dataset by deforming the images with a set of smooth deformations.

Acknowledgements. This work was supported by Microsoft Research through its PhD Scholarship Programme and by the European Research Council through the ERC Advanced Grant MedYMA 2011-291080 (on Biophysical modelling and Analysis of Dynamic Medical Images).

References

1. Kutra, D., Saalbach, A., Lehmann, H., Groth, A., Dries, S.P.M., Krueger, M.W., Dössel, O., Weese, J.: Automatic multi-model-based segmentation of the left atrium in cardiac MRI scans. In: Ayache, N., Delingette, H., Golland, P., Mori, K. (eds.) MICCAI 2012, Part II. LNCS, vol. 7511, pp. 1–8. Springer, Heidelberg (2012)
2. Ecabert, O., Peters, J., Schramm, H., Lorenz, C., von Berg, J., Walker, M.J., Vembar, M., Olszewski, M.E., Subramanyan, K., Lavi, G., Weese, J.: Automatic model-based segmentation of the heart in CT images. IEEE Transactions on Medical Imaging 27(9), 1189–1201 (2008)

3. Karim, R., Mohiaddin, R., Rueckert, D.: Automatic Segmentation of the Left Atrium. In: Medical Image Under standing and Analysis Conference (2007)
4. Depa, M., Sabuncu, M.R., Holmvang, G., Nezafat, R., Schmidt, E.J., Golland, P.: Robust Atlas-Based Segmentation of Highly Variable Anatomy: Left Atrium Segmentation. In: Camara, O., Pop, M., Rhode, K., Sermesant, M., Smith, N., Young, A. (eds.) STACOM 2010. LNCS, vol. 6364, pp. 85–94. Springer, Heidelberg (2010)
5. Margeta, J., Geremia, E., Criminisi, A., Ayache, N.: Layered Spatio-temporal Forests for Left Ventricle Segmentation from 4D Cardiac MRI Data. In: Camara, O., Konukoglu, E., Pop, M., Rhode, K., Sermesant, M., Young, A. (eds.) STACOM 2011. LNCS, vol. 7085, pp. 109–119. Springer, Heidelberg (2012)
6. Lempitsky, V., Verhoek, M., Noble, J., Blake, A.: Random Forest Classification for Automatic Delineation of Myocardium in Real-Time 3D Echocardiography. In: Ayache, N., Delingette, H., Sermesant, M. (eds.) FIMH 2009. LNCS, vol. 5528, pp. 447–456. Springer, Heidelberg (2009)
7. Geremia, E., Clatz, O., Menze, B.H., Konukoglu, E., Criminisi, A., Ayache, N.: Spatial decision forests for MS lesion segmentation in multi-channel magnetic resonance images. NeuroImage 57(2), 378–390 (2011)
8. Criminisi, A., Shotton, J., Konukoglu, E.: Decision Forests: A Unified Framework for Classification, Regression, Density Estimation, Manifold Learning and Semi-Supervised Learning. Foundations and Trends in Computer Graphics and Vision 7(2-3), 81–227 (2011)
9. Tobon-Gomez, C., Peters, J., Weese, J., Pinto, K., Karim, R., Schaeffter, T., Razavi, R., Rhode, K.S.: Left Atrial Segmentation Challenge: a unified benchmarking framework. In: Statistical Atlases and Computational Models of the Heart (2013)
10. Nyúl, L.G., Udupa, J.K.: On standardizing the MR image intensity scale. Magnetic Resonance in Medicine 42(6), 1072–1081 (1999)
11. Otsu, N.: A threshold selection method from gray-level histograms. IEEE Transactions on Systems, Man and Cybernetics C(1), 62–66 (1979)
12. Sato, Y., Nakajima, S., Atsumi, H., Koller, T., Gerig, G., Yoshida, S., Kikinis, R.: 3D multi-scale line filter for segmentation and visualization of curvilinear structures in medical images. In: Troccaz, J., Mösges, R., Grimson, W.E.L. (eds.) CVRMed-MRCAS 1997. LNCS, vol. 1205, pp. 213–222. Springer, Heidelberg (1997)

Multiscale Study on Hemodynamics in Patient-Specific Thoracic Aortic Coarctation

Xi Zhao[1], Youjun Liu[1], Jinli Ding[1], Mingzi Zhang[2], Wenyu Fu[3,4], Fan Bai[1],
Xiaochen Ren[1], and Aike Qiao[1,*]

[1] College of Life Science and Bio-engineering, Beijing University of Technology,
No. 100 Pingleyuan, Chaoyang District, Beijing, China. 100124
[2] Graduate School of Engineering, Tohoku University, Ohta Laboratory,
Institute of Fluid Science, 2-1-1 Katahira Aoba-ku Sendai Miyagi, Japan, 980-8577
[3] College of Mechanical and Electrical Engineering, Beijing Union University,
Beijing 100020, China
[4] College of Architecture and Civil Engineering, Beijing University of Technology,
Beijing 100124, China
qak@bjut.edu.cn

Abstract. In this challenge, we intended to mimic the patient's cardiovascular system by using 0D-3D connected multiscale model. The purpose of the multiscale analysis is to find out the appropriate boundary conditions of the innominate artery (IA), left common carotid artery (LCA) and left subclavian artery (LSA) in the local 3D computational fluid dynamics simulation. Firstly, a lumped parameter model(LPM) of the patient's circulatory system was established which could mimic both the rest and stress conditions by adjusting parameters like elastance function of the heart and the peripheral resistance, since that administering is oprenaline leads to the patient's heart beat rate and peripheral resistance changes. Secondly, the values of parameters in the LPM were slightly revised to match the following conditions: 1. provided pressure and flow rate curves, 2. provided blood distribution ratio of the AcsAo, IA, LCA and LSA. Finally, we got the outlet conditions of the IA, LCA and LSA, and then connecting the 0D model and the 3D model at each time step. As the results, we got the streamlines, pressure drop through the coarctation, pressure gradient, and some other parameters by coupled multiscale simuation.

Keywords: Thoracic aortic coarctation, Hemodynamics, Computational fluid dynamics, Multiscale simulation, Pressure gradient.

1 Introduction

Coarctation of the aorta (CoA) usually occurs in the thoracic segments of the aorta which often leads to hypertension. A lot of researches have been performed to study the hemodynamic parameters such as the flow velocity and wall pressure of the morbid aorta which have shown that those parameters are closely related to vascular geometry. Therefore, hemodynamic simulation can be performed and then applied to

---

* Corresponding author.

O. Camara et al. (Eds.): STACOM 2013, LNCS 8330, pp. 57–64, 2014.

predict the fluid field through a thoracic aorta in presence of the coarctation. In this paper, a lumped parameter model (LPM)[1] of circulatory system was firstly established in order to evaluate the hemodynamic behaviors inside a thoracic aortic coarctation model which is in lack of enough boundary conditions. In accordance with the provided information, the values of parameters in the LPM were properly adjusted to fit both the rest and stress conditions.

Two schemes were planned to get the pressure gradient through the coarctation. One option is to carry out the stand-alone 3D simulation by setting the LPM results as boundary conditions. The other option is to perform multiscale simulation by coupling the LPM and the local 3D model at each time step[2], which is regarded to be able to better reproduce the boundary conditions and represent the interactions between the local geometry and the global circulatory system. As the first results, we performed the stand-alone 3D simulation by setting the 0D simulation results (the volume flow rate curves at the outlets of IA, LCA and LSA) as the boundary conditions which are not given. Then, we performed multiscale simulation to obtain the coupled solutions.

2   Method

(a) Construction of the lumped parameter model

The LPM of the circulatory system was constructed as shown in Fig. 1. In accordance with the given waveforms and flow distribution ratio of the IA, LCA, LSA and DesAo, the values of parameters in the model were properly adjusted to fit both the rest and stress conditions, so that the waveforms of the ascending aortic flow rate, pressure and blood distribution ratio are similar to the curves and values measured from the clinical practice. The comparisons between the original waveforms and LPM waveforms are depicted in Fig. 2, and Table 1 gives the values of parameters in the LPM for both the rest and the stress conditions.

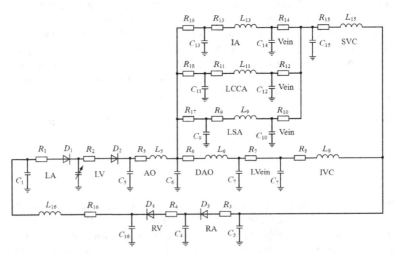

Fig. 1. The LPM of circulatory system (AO: aorta, DAO: descending aorta, IVC: inferior vena cava, SVC: superior vena cava, LA: left atrium, RA: right atrium, LV: left ventricle, RV: right ventricle)

Table 1. The values of Parameters in the LPM for the rest and stress conditions

R(mmHg·s·ml^{-1})			C(ml^{-1}mmHg)			L(mmHg·s^{2}·ml^{-1})		
	Rest	*Stress*		*Rest*	*Stress*		*Rest*	*Stress*
R1	0.01	0.0091	C1	0.2	1.6			
R2	0.01	0.0091						
R3	0.01	0.004	C3	4.8	4.74			
R4	0.001	0.004	C4	4.2	4.78			
R5	0.0001	0.1172	C5	0.82	0.028	L5	3.0E-3	2.0E-5
R6	0.06	0.0352	C6	0.014	0.58	L6	1.75E-3	1.5E-5
R7	0.45	0.0105	C7	1.71	0.081			
R8	0.8	0.008	C8	0.505	0.0105	L8	5.0E-4	5.0E-5
R9	1.799	0.192	C9	0.1	0.01	L9	5.0E-4	5.0E-4
R10	0.385	0.0185	C10	0.01	0.001			
R11	2.439	0.514	C11	0.1	0.01	L11	5.0E-4	5.0E-4
R12	0.385	0.0185	C12	0.01	0.001			
R13	0.87	0.027	C13	0.1	0.01	L13	5.0E-4	5.0E-4
R14	0.485	0.0185	C14	0.01	0.001			
R15	1.735	0.001735	C15	0.505	0.105	L15	5.0E-4	5.0E-4
R16	0.0105	0.02	C16	0.2	3.08	L16	5.0E-4	5.0E-4
R17	0.02	0.08						
R18	0.018	0.08						
R19	0.019	0.08						

Fig. 2. The comparisons between the original waveforms and the LPM results

The purpose of setting up the LPM is to obtain the unknown boundary conditions at the outlets of IA, LCA and LSA. Once the total flow and flow distribution ratio of each outlet under both rest and stress conditions could match the given values, at the same time the waveforms of ascending aortic flow rate and pressure could be similar to the provided curves, then we assume that the LPM could mimic the patient's cardiovascular system to some extent.

Therefore, the LPM could be coupled with the 3D model. The provided inflow waveform was added as inlet boundary condition. The flow rate waveforms of IA, LCA, LSA and the pressure of descending aorta were calculated from the multiscale simulation.

The structure of the coupled multiscale model is shown in Fig. 3. The flow rate of IA, LCA, LSA and the pressure of descending aorta generated by the LPM are set as boundary conditions, and the pressure of IA, LCA, LSA and the flow rate of the descending aorta calculated by the 3D model are passed to the LPM for the calculation at next time step. The data exchange is executed in every time step.

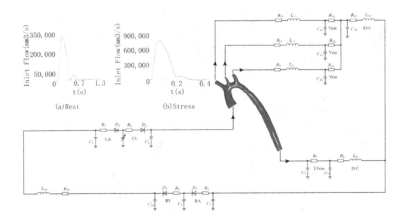

Fig. 3. The structure of the coupled model and the inlet boundary condition

(b) Construction of the finite element model

The provided STL file of arterial model was imported into ANSYS ICEMCFD13.0. Volume meshes were generated by using mesh types of structural hexahedral. The boundary layer was not treated specially. The total numbers of the element and node are 617786 and 480032 respectively. The mesh of a cross-section is shown in Fig. 4.

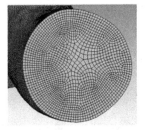

Fig. 4. The mesh in a cross-section

(c) Numerical simulation

Volume mesh file was imported into ANSYS CFX 13.0 to perform the numerical simulation. The following assumptions were employed in this numerical study: non-permeability, rigid wall; incompressible Newtonian fluid; pulsatile and laminar flow. Viscosity and density of blood are 0.004Pa•s and 1000kg/m^3 respectively.

The discrete form of differential equations governing the blood flow was upwind scheme. Residual convergence criteria of mass and momentum were set to 10^{-5}. The time step in calculation was 0.005s. Run mode in CFX is "PVM Local Parallel". Convergent solutions were obtained after 3 cycles.

3 Result

(a) The flow rates and pressure of IA, LCA and LSA
The flow rates of IA, LCA and LSA are set as the boundary conditions in the 3D model, and the pressures of them are the input of the LPM. Both of them can only be determined after the multiscale simulation. All of them are shown in Fig. 5.

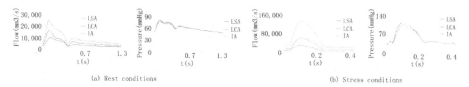

(a) Rest conditions   (b) Stress conditions

Fig. 5. The flow rates and pressure of IA, LCA and LSA

(b) Pressure drop and pressure gradient
As shown in Table 2, both the peak and time-averaged pressure drop and volume-average pressure gradient through the coarctation at both rest and stress conditions (the planes for calculating the pressure drop and the pressure gradient are depicted in Fig. 6) were obtained.

Table 2. Both the peak and time-averaged pressure drop and volume-averaged pressure gradient

	Pressure Drop(mmHg)		Volume-averaged Pressure Gradient(Pa/m)	
	Peak	Time-averaged	Peak	Time-averaged
Rest conditions	5.69	0.1094	135799	22869.5
Stress conditions	26.034	-0.53663	275898	91339.49

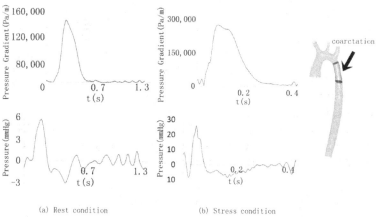

(a) Rest condition   (b) Stress condition

Fig. 6. The pressure drop and volume-averaged pressure gradient through the coarctation and the location of the coarctation

(c) The flow distribution ratio and the pressure proximal to the coarctation

The values of parameters in the LPM were properly adjusted in order to match both the total flow and percentage of ascending aortic flow through the various branches under both rest and stress conditions. As a reference for comparison, table 3 gives the specific distribution values of each opening.

Table 3. The flow distribution ratio and total flow of the various branched in the aortic model under both rest and stress conditions

			AscAo	Inno-minate	LCC	LS	Dia-phAo
Rest Condi-tions	calcul ated	Total Flow(L/min)	3.71	0.607	0.287	0.397	2.419
		% AscAo	100	16	8	11	65
	pro-vided	Total Flow(L/min)	3.71	0.624	0.312	0.364	2.41
		% AscAo	100	17	8	10	65
Stress Condi-tions	calcul ated	Total Flow(L/min)	13.53	3.277	0.6785	1.4901	8.084
		% AscAo	100	24	5	11	60
	pro-vided	Total Flow(L/min)	13.53	3.355	0.6875	1.4575	8.03
		% AscAo	100	25	5	11	59

The systolic, diastolic, and mean pressures of the ascending aorta were measured from the CFD results. Table 4 gives those values which are very close to the value measured clinically.

Table 4. The systolic, diastolic, and mean pressure proximal to the coarctation

Pressure(mmHg)		Systolic	Diastolic	Mean
Rest Conditions	calculated	83.98	49.80	63.36
	provided	83.92	49.68	63.35
Stress Conditions	calculated	118.44	36.38	61.75
	provided	123.35	36.77	64.30

(d) Streamlines, Pressure and Pressure Gradient

The typical moments of 0.255s, 0.83s (under rest condition) and 0.095s, 0.42s (under stress condition) which were the highest and lowest velocity peak point respectively were selected to demonstrate the 3D simulation results.

(i) Streamlines

Figure 7 shows the streamlines at the time of maximum velocity under both the rest and the stress conditions. Results show that the maximum velocity region appears through the coarctation in both conditions.

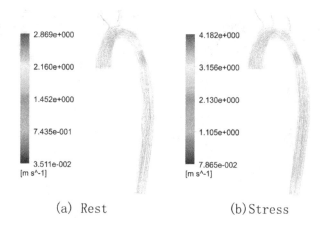

(a) Rest   (b) Stress

Fig. 7. The streamline at the time of maximum velocity in rest and stress condition

(ii) Pressure and Pressure Gradient

The contours of pressure and pressure gradient are showed in Fig. 8 and Fig. 9.

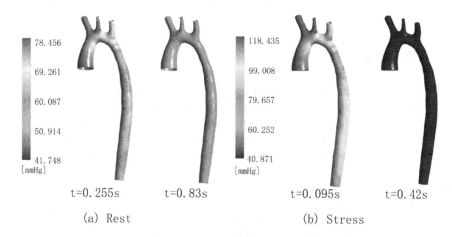

t=0. 255s   t=0. 83s   t=0. 095s   t=0. 42s

(a) Rest   (b) Stress

Fig. 8. The pressure at two typical moments under both rest and stress conditions

It can be seen from the pressure contour that, at the time of highest peak velocity, the pressure difference between the coarctation area and the regions before or after the area is dramatic. However in the case of lowest peak velocity, the pressure difference is not that much.

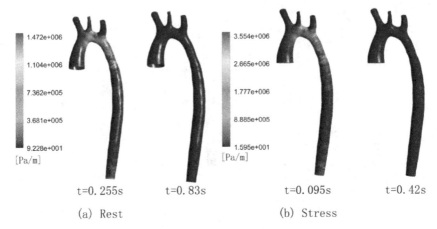

Fig. 9. The pressure gradient at two typical moments under both rest and stress conditions

According to the contour of pressure gradient distribution, it can be found that the pressure gradient through the coarctation is relatively high. However, the highest pressure gradient is located at the bottom of the aortic arch, and the pressure gradient at the time of minimum velocity is much lower than that at the time of maximum velocity.

4   Conclusion

In this paper, the multiscale simulation was presented by coupling the LPM and the 3D model, which is able to better reproduce the boundary conditions and represent the interactions between the local geometry and the global circulatory system.

The results showed that the flow distribution through the various branches and the proximal systolic, diastolic, and mean pressure match the provided data very well.

Acknowledgements. This work was supported by National Natural Science Foundation of China (81171107, 11172016, 10972016), Higher School Specialized Research Fund for the doctoral program funding issue (20111103110012) and the Natural Science Foundation of Beijing (3092004, KZ201210005006).

References

1. Migliavacca, F., Dubini, G., Pennati, G., Pietrabissa, R., Fumero, R., Hsia, T.Y., de Leval, M.R.: Computational model of the fluid dynamics in systemic-to-pulmonary shunts. J. Biomech. 33(5), 549–557 (2000)
2. Moghadam, M.E., Vignon-Clementel, I.E., Figliola, R., Marsden, A.L.: A modular numerical method for implicit 0D/3D coupling in cardiovascular finite element simulations. J. Comput. Phys. 244, 63–79 (2013)

Hemodynamic in Aortic Coarctation Using MRI-Based Inflow Condition

Jens Schaller[1], Leonid Goubergrits[1], Pavlo Yevtushenko[1], Ulrich Kertzscher[1],
Eugénie Riesenkampff[2], and Titus Kuehne[3]

[1] Biofluid Mechanics Laboratory, Charité - Universitätsmedizin Berlin, Germany
{jens.schaller,leonid.goubergrits,pavlo.yevtushenko}@charite.de
[2] Department of Congenital Heart Disease and Paediatric Cardiology, Deutsches
Herzzentrum Berlin, Germany
riesenkampff@dhzb.de
[3] Non-invasive Cardiac Imaging in Congenital Heart Disease Unit,
Charité-Universitätsmedizin Berlin and German Heart Institute Berlin, Germany
titus.kuehne@charite.de

Abstract. Image-based CFD can support diagnosis, treatment decision
and planning. The ability of CFD to calculate pressure drop across the
aortic coarctation is the focus of the 2013 STACOM Challenge. The
focus of our study was inflow conditions. We compared a MRI-based
inlet velocity profile with a swirl and an often used plug velocity profile
without swirl. The unsteady flow simulations were performed using the
solver FLUENT with consideration of the challenge specifications. For
outflows, the constant outflow ratios of the supra-aortic vessels were set.
The consideration of a secondary flow (swirl) at the inlet of the ascending
aorta significantly affect (reduce) the calculated pressure drop across
the aortic coarctation and hence the treatment decision. Furthermore,
using MRI-measured flow rates at the ascending and the descending aorta
without a proof of data consistency could result in an overestimated
pressure drop due to overestimated flow into the supra-aortic vessels.

Keywords: CFD, aortic coarctations, pressure drop, MRI-based inflow.

1 Introduction

Aortic-coarctation is a congenital disease of the aorta causing a raised blood
pressure in the upper and a decreased pressure in the lower body. Guidelines
recommend a surgical or catheter-based (stent placement or balloon angioplasty)
treatment when the systolic pressure gradient exceeds 20 mmHg at rest [1]. In
clinical practice, the pressure drop is measured with a catheterization procedure, which is an invasive, uncomfortable for the patients, and associated with a
radiation. Other methods, such as, Doppler echocardiography, two-dimensional
velocity encoded magnetic resonance imaging (VENC-MRI), solving of Pressure-
Poisson or Navier-Stokes equations using 4D VENC-MRI velocity data or image-
based computational fluid dynamics (CFD) to estimate the pressure drop from

O. Camara et al. (Eds.): STACOM 2013, LNCS 8330, pp. 65–73, 2014.
© Springer-Verlag Berlin Heidelberg 2014

imaging data were proposed. However, they did not replace the catheterization procedure until now due to a lack of methodological and clinical validation. Imaging data in combination with CFD have the potential to replace this invasive pressure measurement. Furthermore, hemodynamic simulations in treatment planning can improve outcome. A proof of the CFD methodology is part of the 2013 STACOM CFD challenge. A major part of the image-based CFD methodology is the selection of inlet boundary conditions. Morbuddici et al. showed [3], that the choice of the inlet velocity profile has a significant impact on the wall shear stress distribution and bulk flow effects. An increase of high WSS areas was found in simulations using MRI-based (4D-VENC-MRI) inlet velocity profile if compared with simulations using simplified profiles. Goubergrits et al. found similar results comparing a plug and 4D-VENC-MRI-based velocity profile for peak systolic steady flow simulations [2]. Additionally, the calculated pressure drop could be significantly decreased when a MRI-based inlet profile is used [2]. In the framework of the 2013 STACOM CFD challenge, the cycle-averaged pressure drop has to be estimated for rest and stress conditions by the proper choice of inflow and outflow conditions. For the inlet at the ascending aorta only the unsteady flow rate measured by 4D-VENC-MRI was provided. In our approach, we used a 3D velocity profile from our data base (4D-VENC-MRI measurements of aortic coarctations) including a secondary flow feature (swirl) at the inlet of the ascending aorta. The profile was adapted to meet the size of the inlet, the anatomical orientation, and the unsteady flow rates of the stressed and rest conditions specified by the challenge.

2   Materials and Methods

2.1   Anatomy Data

The Challenge provided an anatomical model of a mild thoracic aortic coarctation (45% degree of stenosis with a minimal diameter of 10 mm) of a 17-year old male (see Fig. 1). This model includes the ascending aorta (AAo) with an inlet diameter of 21 mm, the arch, supra-aortic-vessels, and descending aorta (DAo) with an outlet diameter of 11.5 mm. The provided surface mesh was smoothed slightly with the Laplace smoothing algorithm using ZIBamira (Zuse-Institute-Berlin, Germany). This surface was imported into the mesh generator Gambit 2.4.6 (Ansys Inc, USA). In a first step all supra-aortic outlets were extended by 10 mm. The surface mesh of these extensions was meshed with 2,300 quadrilaterals and the original part was remeshed with 107,935 triangles. The volume mesh consisted out of two boundary layers at the wall with 211,270 wedge and 4,600 hexahedral elements and the remaining volume was filled with 1,226,734 tetrahedral elements. Cell skewness (9 cells were > 0.75; worst cell: 0.81) and aspect ratio (all cells below 4) meet the flow solver requirements. According our earlier mesh independence study with another aortic coarctation case, the used mesh allows a calculation of velocity and pressure fields within 1% error [2].

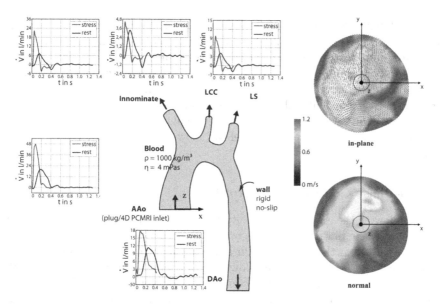

Fig. 1. Boundary conditions. Left: Used flow rates and boundary conditions for the rest and stress (LCC - left common carotid artery, LS - left subclavian artery). Right: MRI-based adapted inflow velocity profile (in-plane and normal velocity components).

2.2 Simulations

With the flow solver Fluent 6.3.26 (Ansys Inc., USA), three-dimensional unsteady laminar flow simulations of the two conditions (rest/stress:3.7/13.5 L/min) were performed. The fluid properties were set to a density of 1000 kg/m^3 and a constant viscosity of 4.0 mPas. Peak systole Reynolds numbers at the inlet were 5,000 (rest) and 12,000 (stress). The peak systole Reynolds numbers at the stenosis were 5,980 and 9,350 respectively. Womersley numbers were 23 (rest) and 40 (stress). The wall was rigid and no-slip condition was set.

To meet the provided unsteady flow rates for the inlet and outlets, user defined functions (UDFs) were written using MATLAB (Mathworks Inc., USA) and applied as velocity inlets in the flow solver (see Fig. 1).The inlet profile was either a plug profile or the adapted 4D-VENC-MRI-based profile described below. When using the plug profile, a constant normal velocity at the inlet was set. At the outlets, which have a sufficient distance to the coarctation, also plug profiles without secondary flow were imposed except the outlet of the Innominate (see Fig. 1). For this outlet the Fluent-implemented outlet boundary condition outflow was used. The flow rate curves of the supra-aortic-vessels were the same for all three branches calculated as a difference between provided flow rate curves for the ascending and descending aortas but with respective scaling fulfilling the Challenge requirements. This implies a fixed flow split ratio between the supra-aortic-vessels. Using this boundary condition setup, only relative pressure fields were calculated. The absolute physiologic pressures can be calculated using the pressure curve at the proximal plane in the AAo as a reference. The flow was solved with double precision, pressure-based solver using 2nd order discretization in space, 2nd order implicit in time, and a SIMPLEC-pressure-velocity-coupling. 1,000 time steps per cycle for the stress and 1,500 time steps per cycle for the rest condition were calculated, to obtain equal numbers of calculated time steps in the systolic phase. The time steps were simulated with up to 120 iteration per time step or up to a residual level of below 5-e5 for x-, y-, z-velocity and continuity.

2.3   3D-Phase Contrast MRI Profile

For the inflow at the AAo, a 3D velocity profile with a moderate secondary flow (swirl) available in our data base was used. The secondary flow in the aorta inlet was characterized as the ratio of the in-plane mean velocity to the through-plane mean velocity (degree of secondary flow - DSF). Here the DSF was equal to 1.08 that is a moderate secondary flow in comparison to the maximal DSF in our data base of 2.5 (zero means no secondary flow). Data were obtained from a 23 years old woman (88 bpm heart rate) with aortic coarctation and a 47% stenosis at the Department of Congenital Heart Disease and Peadiatric Cardiology at the German Heart Institute in Berlin, Germany using a 1.5 Tesla MR scanner. The velocity was resolved with a spatial resolution of 1.7 mm x 1.7 mm x 2.5 mm. The temporal resolution was 27 ms. The peak systolic velocity profile (see Fig. 1) was extracted from the 4D-VENC-MRI data with MEVISFlow 8 (Fraunhofer MEVIS, Bremen, Germany) and used in this study as a constant inlet velocity profile scaled for each time step according the inlet flow rate curve. We paid attention to extract the velocity profile from a similar anatomical position and orientation: Both aorta surfaces were aligned manually to achieve a good match of the AAo's centerlines. The velocity profile was extracted from cross-section that was normal to the vessel centerline.

3   Results

Fig. 2 shows pressure drops at cross section defined by the STACOM Challenge for the rest and stress conditions simulated with plug and MRI-based inlet velocity profiles. By using the same flow rate, the secondary flow in the inlet of the AAo described by MRI-based inlet velocity profile results in a lower maximal pressure drop for both flow conditions (rest and stress): 18.3 mmHg vs. 22.0 mmHg during the rest and 65.6 mmHg vs. 91.3 mmHg at stress. Cycle averaged pressure drops were also significantly ($p<0.05$) reduced in MRI-based simulations: rest 2.1 ± 5.29 mmHg vs. 2.9 ± 6.65 mmHg; stress 4.7 ± 16.86 mmHg vs. 10.9 ± 23.03 mmHg (paired Wilcoxon test). Differences of the flow fields between plug and MRI-based conditions causing lower pressure drop in MRI-based simulations are shown in Fig. 5. A strong secondary flow structure (swirl) is formed in the AAo of the MRI-based simulations. Note that peaks of pressure

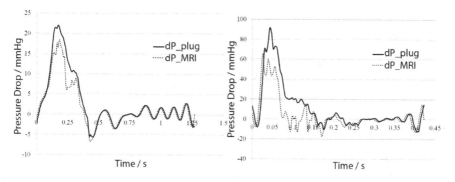

Fig. 2. Time courses of the pressure drop calculated for the rest (left) and stress (right) conditions with plug and MRI-based inlet velocity profiles

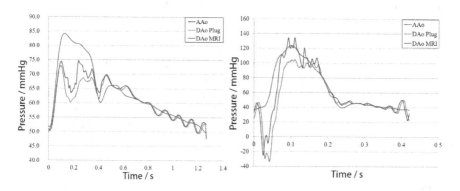

Fig. 3. Time courses of the space-averaged static pressure at STACOM-specified cross sections in AAo and DAo calculated for the rest (left) and stress (right) conditions with plug and MRI-based inlet velocity profiles

drops are calculated at time points (0.17 s – rest; 0.038 s – stress, see Fig. 2), which do not correlate with peak flow through the aortic coarctation (0.25 s – rest; 0.08 s – stress, see Fig. 1).

Using the calculated pressure drops and the provided pressure curve at the AAo, the pressure at the DAo is calculated (see Fig. 3). The curves of MRI and plug inlet show common course. However, in the stress condition the curves show clearly an unphysiological negative pressure at the beginning of systole.

Fig. 4 shows pressure decrease downstream from the inlet at peak systolic phase. In both pressure curves, three common local minima and maxima were found at sites A, B and C. From a certain site D downstream a monotonic decrease is seen. Between C and D there is an increase of static pressure that is associated with filling of the flow at the aortic cross-section beyond the coarctation.

The visualization of a pressure field cut reveals the origin of the progression: A low pressure region occurs at the inner aortic arch even before the branch of the Innominate artery (see Fig.5). The pressure field of the rest condition shows a more continuously pressure decrease in comparison to the stress condition. In the

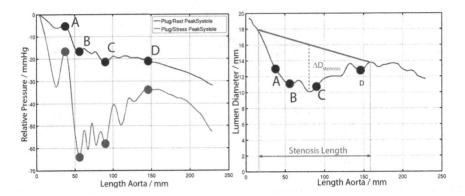

Fig. 4. Left - Relative static pressure local minus inlet static pressure difference along the aorta for plug simulations at peak systole. The markers A-B-C-D were set 36/55/90/146 mm away from beginning of the centerline.

stress case, a large drop is seen in the aortic arch (up to 100 mm from the inlet) but further downstream the pressure recovers of about 25 mmHg and reduces the overall pressure drop. This low pressure region correlates with a recirculation zone that can be clearly seen in Fig. 5. The site A coincides with the beginning of the recirculation zone. At site B, the recirculation zone has its largest spatial extend inside the lumen. The streamlines of the rest case are aligned earlier than in the stress case.

4   Discussion

We investigated two different in–flow conditions: use of the plug inlet velocity profile with the MRI-measured flow rate provided by the STACOM Challenge and use of an MRI-based inlet velocity profile taken from our own data base. At the outlets we used simple boundary conditions without the often used approach of a 1D three-element Windkessel modeling. This is motivated by the fact that Windkessel modeling requires a choice of 9 unknown and patient/nonspecific parameters for the three supra-aortic-vessels, when the flow rates at the AAo and DAo are described by the MRI-provided flow rate curves. On the other hand the aortic coarctation of the Challenge patient is located downstream of the aortic arch and flow rate division among side branches does not affect the flow rate through the aortic coarctation that is equal to the flow rate of the descending aorta provided by the STACOM Challenge. Taking into account studies [4], which found that comprehensive 1D personalized pressure-drop models are able to predict pressure drop due to aortic coarctation, we do not expect that a complex variability of flow rate divisions affect aortic coarctation pressure drop in our case. The calculated pressure drop between AAo and DAo of about 20 mmHg at rest condition assign the coarctation as a borderline case of a treatment decision. Our results comparing simulations with plug and MRI-based velocity profile illustrate the huge meaning of inlet boundary conditions.

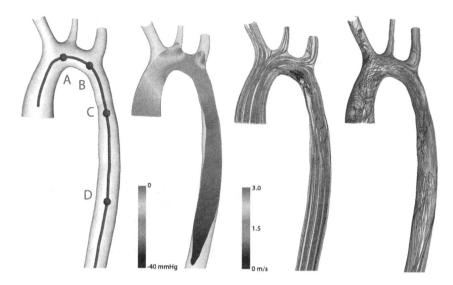

Fig. 5. Flow and pressure fields at rest. From left to right: Centerline with landmarks A-D. Cut plane with relative static pressure at AAo peak flow and plug inlet profile. Respective streamlines with color-coded velocity magnitude. Streamlines with color-coded velocity magnitude at AAo peak flow and MRI-based inlet profile.

Based on plug-simulation at the rest condition the patient should be treated (maximal dP=22 mmHg), whereas MRI-based simulation could recommend no treatment (maximal dP=18.3 mmHg). Lower pressure drops in simulations of aortic coarctations using MRI/based inlet velocity profile correlate well with our recent study with 3 other aortic coarctation cases and with a statement of Kilner et al. that helical flow in the aorta is a part of the natural flow optimization process [5].

The calculated absolute pressure value range especially in the stressed condition has unrealistic low pressure values. Using the provided time series of flow rates at the AAo and DAo neglect the fact that the values were averaged, and are not simultaneously acquired data. The problem is emphasized in Fig. 6 for both the rest and the stress condition. The difference between the two flow rate curves is the flow rate into the supra-aortic-vessels (green solid lines). Especially at the rest condition the resulting curve with about 250 ml/s maximal flow in upper body and about 120 ml/s maximal reversed flow shows non-physiologic flow conditions in the upper body for rest conditions. Using a relatively low time shift of the DAo flow curve by 0.06 s results in a physiological upper body flow curve with the maximal flow rate of about 140 ml/s and a low reversed flow (see Fig. 6). Similarly, for the stress condition the shift of the flow curve by 0.01 s reduces maximal flow rate to the upper body from 600ml/s to 500 ml/s. The simulated high maximal flow rates into the supra-aortic-vessels correlate temporally with the maximal pressure drop between AAo and DAo. This is the consequence of the high flow into to the supra-aortic-vessels, which needs a high

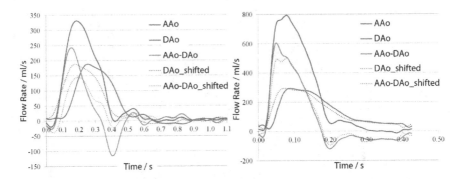

Fig. 6. Left: Flow rates measured by 2D-VENC-MRI for the ascending (total flow) and descending (flow into the lower body) aortas and their difference (solid lines) , which is the flow into the supra-aortic-vessels, for rest condition. Dotted lines show descending aorta flow shifted in time by dt = 0.06 s and corresponding flow into the supra-aortic-vessels. Right: Flow rates for stress condition. The flow rate of the descending aorta was shifted by dt = 0.01 s.

pressure difference between the AAo and the supra-aortic-vessels. This defines the pressure level in the aortic arch upstream of the coarctation. An additional pressure drop result from the coarctation, which is defined by the flow rate through it. Both pressure drops together results in the pressure drop between AAo and DAo. Note that a total cross-section area of the three side branches with 96 mm² is of the same order as the cross-section area of the descending aorta outlet with 112 mm² and of the aortic coarctation (100 mm²) and therefore both pressure drops are from the same order. As the clinical main concern is the part of the pressure drop caused by the coarctation, it must be separated from the total pressure drop.

The provided flow rate curves neglect, however, the unknown storage capacity of blood due to Windkessel effect of the aorta during systole. This can also play a role to compensate the disadvantageous effects.

5   Conclusion

Aortic coarctation pressure drop assessment using image-based CFD is very sensitive to the inlet velocity profile. The modeling of the secondary flow feature (swirl) generated by the left ventricle together with the aortic valve at the ascending aorta affect pressure drop calculation that could influence treatment decision. This is especially important for borderline patients. Furthermore, attention should be paid to the MRI-measured flow rate curves at the ascending and descending aortas, which are used as boundary conditions of CFD simulations. These data are not simultaneously acquired data meaning that a difference of flow rate curves could result in a non-physiologic flow rate in the supra-aortic-vessels. Consequently a use of 1D circulation models validating MRI-measured data including secondary swirl could be recommended.

References

1. Warnes, C., Williams, R., Bashore, T., Child, J., Connolly, H., Dearani, J., del Nido, P., Fasules, J., Graham, T., Hijazi, Z., Hunt, S., King, M., Landzberg, M., Miner, P., Radford, M., Walsh, E., Webb, G.: ACC/AHA, Guidelines for the Management of Adults With Congenital Heart Disease. Circulation 118, E714–E833 (2008)
2. Goubergrits, L., Mevert, R., Yevtushenko, P., Schaller, J., Kertzscher, U., Meyer, S., Schubert, S., Riesenkampff, E., Kuehne, T.: The Impact of MRI-Based Inflow for the Hemodynamic Evaluation of Aortic Coarctation. Ann. Biomed. Eng. (in press)
3. Morbiducci, U., Ponzini, R., Gallo, D., Bignardi, C., Rizzo, G.: Inflow boundary conditions for image-based computational hemodynamics: impact of idealized versus measured velocity profiles in the human aorta. J. Biomech. 46, 102–109 (2013)
4. Itu, L., Sharma, P., Ralovich, K., Mihalef, V., Ionasec, R., Everett, A., Ringel, R., Kamen, A., Comaniciu, D.: Non-invasive hemodynamic assessment of aortic coarctation: validation with in vivo measurements. Ann. Biomed. Eng. 41, 669–681 (2013)
5. Kilner, P.J., Yang, G.Z., Mohiaddin, R.H., Firmin, D.N., Longmore, D.B.: Helical and retrograde secondary flow patterns in the aortic arch studied by three-directional magnetic resonance velocity mapping. Circulation 88, 2235–2247 (1993)

Sensitivity Analysis of the Boundary Conditions in Simulations of the Flow in an Aortic Coarctation under Rest and Stress Conditions

Salvatore Cito*, Jordi Pallarés, and Anton Vernet

Universitat Rovira i Virgili, Department of Mechanical Engineering,
Avinguda dels Paisos Catalans, 26, Campus Sescelades,
43007, Tarragona, Spain
salvatore.cito@urv.cat

Abstract. Aortic coarctation is a congenital condition in which the aorta is narrowed. The pressure drop through the aorta depends on the extension of this narrowing. Therefore, severe coarctations have important side-effects on the hemodynamic conditions of the circulatory system. The Computational Fluid Dynamics (CFD) simulation of the hemodynamics in a patient specific aortic coarctation can help to predict the pressure drop produced by the aorta narrowing. Nevertheless, an accurate prediction of the pressure drop, by means of CFD tools, depends strongly on the boundary conditions (BCs). In this study we present the results of ten simulations performed on a patient specific aortic coarctation vasculature proposed in the second MICCAI-STACOM CFD Challenge. The model includes ascending aorta, arch, descending aorta, and upper branch vessels. Specifically, we consider; two different physiological states a) rest and b) stress condition. For both cases the flow rate waveforms are provided at the ascending aorta and descending aorta. The ten simulations are performed using five different set of boundary conditions. The results are reported and indications on the more accurate boundary conditions are discussed.

1 Introduction

Aortic coarctation can have important side-effects on the hemodynamic conditions of the circulatory system and can lead to hypertension [1]. Advances in patient specific computational hemodynamic techniques allow to simulate blood flow and pressure in thoracic coarctation models extracted from patient data. These emergent technologies offer the possibility of predicting the pressure drop through the coarctation [2]. This can be satisfactory accomplished if; 1) measurements of the flow through the main vessel branching from the aorta are available 2) the three dimensional virtual phantom of the aorta is accurately reproduced and 3) information about vessel thickness and elasticity are available. Unfortunately, in the clinical practice only part of this information can

---

* Corresponding author.

O. Camara et al. (Eds.): STACOM 2013, LNCS 8330, pp. 74–82, 2014.

be obtained. Consequently, this lack of information can lead to consider a wide range of different boundary conditions for the simulation. This can affect the consistency and repeatability of the simulations. The objective of this study is to assess the predictive power of CFD methods in computing the blood pressure drop through a moderate aortic coarctation model, under rest and stress conditions. The case under study has been set within the STACOM 2013 CFD challenge [3]. Specifically five different set of boundary condition are used and their effect on the pressure at rest and stress is discussed.

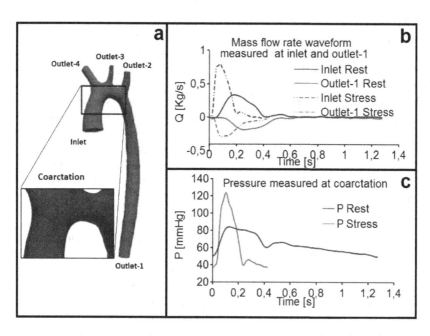

Fig. 1. a) Rendering of MRA and stl file representing the aorta with a close-up view of the surface mesh. b) Ascending aortic and diaphragmatic aortic flow waveforms at rest condition and under stress conditions. c) Pressure wave at coarctation.

2 Material and Methods

The anatomic data and of a moderate aortic coarctation at rest and the physiological flow condition at rest and under stress conditions of the subject of this study are provided by the organizers of the STACOM 2013 CFD challenge [3]. Figure 1.a shows the virtual phantom of the segmentation from the MRA data. The model includes the ascending aorta, arch, descending aorta, and upper branch vessels. The cardiac output of the patient at rest was 3.71 L/min, the heart rate 47 beats per minute (cardiac cycle $T = 1.277$ s). The cardiac output of the patient, under stress condition, increased to 13.53 L/min, the heart rate to 141 beats per minute (cardiac cycle $T = 0.425 \, s$). The proximal systolic, diastolic, and mean pressures were at rest condition 83.92, 49.68, and 63.35 $mmHg$, respectively, and under stress condition 123.35, 36.77, and 64.30 $mmHg$, respectively [3]. For both physiological conditions (rest and stress), the flow waveforms and the pressure wave were reconstructed at the levels of the ascending aorta (AscAo) and diaphragmatic aorta (DiaphAo). These waveforms are depicted in Figures 1.b and 1.c. The organisers of the challenge provided a 15-term Fourier reconstruction of the flow waveforms [3]. Note that in Figure 1.b the positive flow rates indicate inflow and negative values correspond to outflow.

CFD Model: The blood flow through the three-dimensional virtual phantom was modeled as an incompressible Newtonian fluid. The governing equations are the unsteady incompressible Navier-Stokes equations, which were solved numerically, together with the corresponding boundary conditions, using the commercial CFD code Ansys Fluent [4]. Due to the lack of information about the vessel wall elasticity and thickness, the vessel walls were assumed to be rigid. A constant viscosity equal to 0.004 $Pa \cdot s$ and a constant density equal to 1000 Kg/m^3 were set. Non-Newtonian effects of blood were neglected. One hundred time-steps per cardiac cycle were used, and the calculations were performed for 2 cardiac cycles. The discretized momentum and pressure equations were solved using second order spatial discretization schemes and an implicit second order discretization scheme was used for the time marching.

Table 1. Combinations of boundary conditions BC1, B2, BC3, BC4 and B5 imposed for the simulations under rest and stress conditions

	Outlet-1	Outlet-2	Outlet-3	Outlet-4	Inlet	Vessel wall
BC1	Outflow	Outflow	Outflow	Outflow	Velocity-plug	No-slip
BC2	Velocity-plug	Outflow	Outflow	Outflow	Velocity-plug	No-slip
BC3	Velocity-plug	Outflow	outflow	Pressure	Velocity-plug	No-slip
BC4	Pressure	Outflow	Outflow	Outflow	Velocity-plug	No-slip
BC5	Velocity plug	Pressure	Pressure	Pressure	Velocity-plug	No-slip

Boundary Conditions: We used the given physiological data to implement five different combination of boundary conditions (BCs). The BCs are summarized in Table 1. In the next we will refer to these five sets of boundary conditions as: BC1 to BC5, respectively. In all the types of BC the following two conditions are maintained 1) no-slip boundary conditions were prescribed at the vessel walls and 2) pulsatile velocity boundary conditions, according to Fig. 1.b, are imposed at the model inlet by using a uniform plug velocity distribution. The outflow BC used by the commercial code assumes zero diffusion flux for all flow variables along the direction perpendicular to the outlet and it adjusts the instantaneous mass flow rate to satisfy the overall mass balance in the computational domain.

This BC is exact for steady fully developed flows. The different set of boundary conditions used in the simulations are:

1) in the set BC1 we imposed, instantaneously, the flow distribution through the various outlets to match the data given in Table 2, which corresponds to the averaged flow rate distribution through the inlets and outlets provided by the organizers of the challenge.

2) in BC2 the outflow condition at the outlet-1 of BC1 was changed by an unsteady plug velocity boundary condition according to the time evolutions of Fig. 1.b. The instantaneous flow rates at outlets 2, 3 and 4 were distributed according to Table 2.

3) in BC3 we changed the condition at the outlet-4 of BC1 by imposing on this face the pulsatile pressure boundary conditions of the pressure wave proximal to the coarctation (Fig. 1.c). The outflow at outlets 2 and 3 are distributed according to Table 2

4) in BC4 we set the pulsatile pressure boundary conditions by imposing the pressure wave proximal to the coarctation at outlet-1. The outflow at outlets 2, 3 and 4 are distributed according to Table 2.

5) in BC5 we imposed the pulsatile pressure boundary condition at outlet-2, 3 and 4.

It should be noted that with all the different sets of BCs the overall mass flow balance is satisfied, instantaneously, within 0.03%.

Table 2. Total flow through the various branches of the aortic model under rest and stress conditions. Positive values indicate inflow and negative values outflow.

	Outlet-1	Outlet-2	Outlet-3	Outlet-4	Inlet
Flow at rest [L/min]	-2.41	-0.364	-0.312	-0.624	3.71
Flow at stress [L/min]	-8.03	-1.4575	-0.6875	-3.355	13.53

Volumetric Mesh: We constructed three different body-fitted tetrahedral volumetric meshes of the given virtual phantom for the CFD model. Mesh-1 has 280596 cells and five prism layers at the vessel wall, Mesh-2 has 515242 cells and three prism layers at the vessel wall and Mesh-3 has 1348866 cells and five prism layers at the vessel wall. All the meshes were constructed using the commercial code ICEM [5]. The CFD model under stress condition with BC1 has been computed using the three meshes. The results of these three simulations have been used to make a mesh dependence analysis. Specifically, we compared the pressure drop at the centerline at the systolic time. The results of this analysis show that the pressure field for the solution with Mesh-2 and Mesh-3 are very similar (3% error for the case under stress condition and at systolic time t=0.08 s). Following this analysis, all the simulations were carried out with Mesh-3.

Table 3. Average mass flow rate at outlets and inlet in Kg/s and average and maximum pressure at coarctation in $mmHg$

	EXP	BC1	BC2	BC3	BC4	BC5
Average mass flow rate under rest, L/min						
at inlet	3.83	3.83	3.83	3.83	3.83	3.83
at outlet-1	-2.49	-2.49	-2.45	-2.45	0.00	-2.46
at outlet-2	-0.38	-0.38	-0.39	-0.77	-1.43	-0.19
at outlet-3	-0.31	-0.31	-0.31	-0.61	-1.14	-0.17
at outlet-4	-0.65	-0.65	-0.67	0.00	-1.26	-1.01
Average mass flow rate under stress, L/min						
at inlet	13.96	13.96	13.96	13.96	13.96	13.96
at outlet-1	-8.24	-8.24	-8.15	-8.15	0.00	-8.15
at outlet-2	-1.54	-1.54	-1.56	-4.01	-3.74	-1.26
at outlet-3	-0.69	-0.70	-0.71	-1.80	-1.70	-0.71
at outlet-4	-3.49	-3.49	-3.54	0.00	-8.51	-3.84
Average pressure in $mmHg$						
at coarctation under rest condition	63.4	52.2	48.4	58.2	62.2	60.0
at coarctation under stress condition	64.3	50.8	41.0	57.1	62.5	77.8
Maximum pressure in $mmHg$						
at coarctation under rest condition	83.9	67.5	61.8	74.2	81.7	106
at coarctation under stress condition	123	102	159	160	118	217

3   Results and Discussions

In this section we present and discuss the effect of the different boundary conditions on (1) the pressure at the coarctation, which was measured experimentally under rest and stress conditions (Fig.1.c) and (2) on the blood flow rates through the different outlets. This information is summarized in Table 3. The mean and maximum values of pressure for each case and for both physiological situations are also included in Table 3.

The numerically predicted time evolutions of the mass flow rates and pressures under rest condition are plotted in Figure 2, while Figure 3 shows the corresponding data under stress conditions. The available experimental measurements are included in these figures for comparison.

The set of boundary conditions BC1 assumes that the instantaneous flow through the different outlets is proportional to the instantaneous flow at the inlet (see for example Fig. 2.a and BC1 in Fig. 2.b). This is not completely true according to Figure 1.b in which it can be seen that the measured mass flow rate through the outlet-1 is not proportional to the flow rate at the inlet during the cardiac cycle. In fact, Figure 1.b indicates that the absolute value of the flow rate through inlet is larger than that at the outlet-1 for $t < 0.18\ s$ under

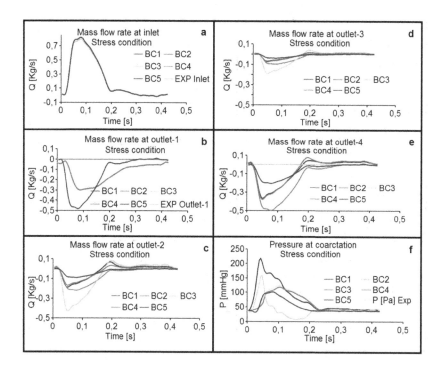

Fig. 2. Time evolutions during a cycle of, (a) to (e), the mass flow rates at the inlets and outlets and (f) of the pressure at the coarctation for the rest conditions. In, (b) to (e), positive values indicate outflow and negative values inflow.

stress conditions and for $t < 0.35\ s$ under rest conditions. This produces that the flow rate through outlets 2, 3 and 4 should exit the computational domain. However at larger times the flow through these surfaces has to enter into the computational domain to satisfy the overall mass balance. These limitations of BC1 produce significant differences between the computed and measured time evolution of the pressure in the coarctation under rest conditions (see Fig. 2.f and Table 3). It is interesting to note that these differences are reduced under stress conditions for $t < 0.07\ s$ (see Fig. 3.f) probably because the experimental distribution of the flow rates through the different branches is well reproduced by BC1 during this period of time.

The sets BC2 and BC3 impose the measured mass flow rate evolution at outlet-1 and the flow rate is distributed through the remaining outlets to satisfy continuity. In BC3 the time evolution of pressure at the coarctation is imposed at outlet-4, which is the closest to the narrowing. Although the flow rate distribution of these sets is closer to the experimental conditions than that of BC1 the numerical predictions of the time evolution of pressure (Figs. 2.f and 3.f and Table 3) using BC2 or BC3 do not improve with respect to BC1. In fact the set

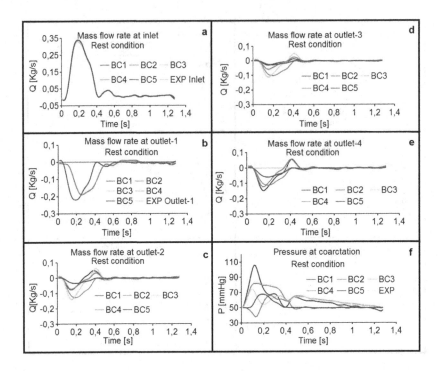

Fig. 3. Time evolutions during a cycle of (a) to (e) the mass flow rates at the inlets and outlets and (f) of the pressure at the coarctation for the stress conditions. In, (b) to (e), positive values indicate outflow and negative values inflow.

BC3 predicts zero flow rate at outlet-4 (Figs. 2.e and 3.e) in which the pressure BC is imposed. The use of the same pressure BC in outlets 2, 3 and 4, as in BC5, overcomes this limitation (see Figs. 2.e and 3.e) but produces significant larger peaks of pressure than those measured (see for example Table 3).

If the time evolution of pressure is imposed at outlet-1 (BC4) the experimental pressure is well reproduced under rest (Fig. 2.f) and stress (Fig. 3.f) conditions but the distribution of flow rates through the different outlets do not agree with that measured (see Table 3). In addition, it can be seen in Figs. 2.b and 3.b that the numerically predicted flow rate through outlet-1 differs from that measured experimentally.

From these considerations, one can conclude that the numerical predictions of the measured time evolution of pressure would need the complete instantaneous time history of the flow rate through all the inlets and outlets during the cardiac cycle instead of the time-averaged distribution available for outlets 2, 3 and 4. Note that even these outlets have a smaller diameter than outlet-1, they, on average, contribute, approximately, 35% and 40% to the total outflow during the cardiac cycle for rest and stress conditions, respectively, as indicated in Table 2. These results suggest that the use of more elaborated boundary conditions, such as reduced-order lumped parameter BC models, need to be considered.

4 Supplementary Material

In order to compare the results of this study to the results of all participants to the Statcom CFD challenge [3], we report the coarctation pressure drops between given proximal and distal planes. The proximal plane correspond to the intersection between the 3D coarctation model and the plane with origin (188.96, 40.18, 253.22) and normal (0.98,-0.09, 0.19). The distal plane correspond to the intersection between the 3D coarctation model and the plane with origin (261.97, 23.56, 277.10) normal (0.99,-0.03, -0.14). These locations correspond roughly to where the invasive pressure wire measurements were acquired. In Figure 4 we report the time-resolved spatial average of the pressure drop between the proximal and distal planes. In Table 4 we report mean and maximum pressure drops (in mmHg) between the proximal and distal locations, respectively under rest and stress condition.

Table 4. Pressure drop between the proximal and distal planes computed using the five different boundary conditions under rest and stress condition

	EXP BC1	BC2	BC3	BC4	BC5
Mean P drop under rest, mmHg	4.01	3.85	4.37	1.71	3.70
Maximum P drop under rest, mmHg	33.33	23.82	35.17	12.87	23.19
Mean P drop under stress, mmHg	21.11	18.74	29.45	9.51	18.14
Maximum P drop under stress, mmHg	212.49	97.11	188.05	46.95	98.10

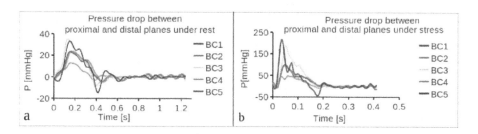

Fig. 4. Time-resolved spatial average of the pressure drop between the proximal and distal planes computed with the five set of boudary conditions. (a) Under rest condition. (b) Under stress condition

References

1. Hager, A.: Hypertension in aortic coarctation. Minerva Cardioangiologica 57(6), 733–742 (2009)
2. Govindaraju, K., Badruddin, I.A., Viswanathan, G.N., Ramesh, S.V., Badarudin, A.: Evaluation of functional severity of coronary artery disease and fluid dynamics' influence on hemodynamic parameters: A review. Physica Medica 29(3), 225–232 (2013)
3. Alberto Figueroa, N.W., Mansi, T., Sharma, P.: 2nd CFD Challenge Predicting Patient-Specific Hemodynamics at Rest and Stress through an Aortic Coarctation (2013), http://www.vascularmodel.org/miccai2013/
4. ANSYS FLUENT 13 (2013), http://www.ansys.com/
5. ANSYS ICEM CFD (2013), http://www.ansys.com/

Patient-Specific Hemodynamic Evaluation of an Aortic Coarctation under Rest and Stress Conditions

Priti G. Albal[1,4], Tyson A. Montidoro[2], Onur Dur[3], and Prahlad G. Menon[1,4,5]

[1] Yat-sen University - Carnegie Mellon University (SYSU-CMU)
Joint Institute of Engineering (JIE), Pittsburgh, PA, USA
[2] Department of Biomedical Engineering, Carnegie Mellon University, Pittsburgh, PA, USA
[3] Thoratec Corporation, Pleasanton, CA, USA
[4] QuantMD LLC, Pittsburgh, PA, USA
[5] SYSU-CMU, Shunde International Joint Research Institute, Guangdong, China
pgmenon@andrew.cmu.edu

Abstract. Computational fluid dynamics (CFD) simulation of internal hemodynamics in complex vascular models can provide accurate estimates of pressure gradients to assist time-critical diagnostics or surgical decisions. Compared to high-fidelity pressure transducers, CFD offers flexibility to analyze baseline hemodynamic characteristics at rest but also under stress conditions without application of pharmacological stress agents which present undesirable side effects. In this study, the variations of pressure gradient and velocity field across a mild thoracic coarctation of aorta (CoA) is studied under pulsatile ascending aortic flow, simulative of both rest and stress cardiac output. Simulations were conducted in FLUENT 14.5 (ANSYS Inc., Canonsburg, PA, USA) - a finite volume solver, COMSOL 4.2a (COMSOL Multiphysics Inc., Burlington, MA) - a finite element solver, and an in-house finite difference cardiovascular flow solver implementing an unsteady artificial compressibility numerical method, each employing second-order spatio-temporal discretization schemes, under assumptions of incompressible, Newtonian fluid domain with rigid, impermeable walls. The cardiac cycle-average pressure drop across the CoA modeled relative to the given pressure data proximal to the CoA is reported and was found to vary significantly between rest and stress conditions. A mean pressure gradient of 2.79 mmHg was observed for the rest case as compared to 17.73 mmHg for the stress case. There was an inter-solver variability of 16.9% in reported mean pressure gradient under rest conditions and 23.71% in reported mean pressure gradient under stress conditions. In order to investigate the effects of the rigid wall assumption, additional simulations were conducted using a 3-element windkessel model implemented at the descending aorta, using FLUENT. Further, to investigate the appropriateness of the inviscid flow assumption in a mild CoA, CFD pressure gradients were also compared results of a simple Bernoulli-based formula, used clinically, using just the peak blood flow velocity measurements (in m/s) obtained distal to the aortic coarctation from CFD. Helicity isocontours were used as a visual metric to characterize pathological hemodynamics in the CoA.

1 Introduction

Coarctation of the aorta (CoA) is a common congenital disease present in adolescence or adulthood, and often identified in the context of investigation for hypertension

O. Camara et al. (Eds.): STACOM 2013, LNCS 8330, pp. 83–93, 2014.
© Springer-Verlag Berlin Heidelberg 2014

[1, 2]. CoA is often associated with other congenital diseases including atrial septa defect, pulmonary stenosis, etc [3, 4]. Decrease in regional diameter at the aortic coarctation results in elevated pressure gradients across it. These pressure gradients increase several fold under stress conditions, in contrast with rest conditions[5]. Compared to high-fidelity pressure transducers employed during invasive pressure measurement, computational fluid dynamics (CFD) offers flexibility to analyze baseline pre-repair hemodynamic characteristics non-invasively at rest and also under stress conditions *in-silico*, without application of pharmacological stress agents which present undesirable side effects, including chest pain, blood pressure decrease, arrhythmia, palpitations, shortness of breath and headaches [6, 7]. CFD can provide information regarding peak velocities distal to CoA and visual representations of flow structures using flow derived parameters such as helicity isocontours which can objectify characterizing the pathological extent of pre-repair CoA hemodynamics[8]. In this study, we use CFD to predict pressure gradients across a mild CoA (~30%) in a17-year old male patient, at both rest (Re = 95) and pharmacological stress conditions (Re = 3400) [5]. We provide a comparison of results between two commercial solvers – a finite volume (FVM) and a finite element (FEM) solver – as well as one in-house finite difference (FDM) solver, each set up to model rigid-wall CFD supplied with the same inflow and outflow conditions specified for the 2013 STACOM CFD Challenge. Pressure gradients from CFD were compared against those obtained from the commonly used clinical formula [9], $\Delta P = 4V^2$, where V is the peak velocity (in m/s) distal to CoA, in order to assess the validity of the Bernoulli assumptions it is based upon (i.e. inviscid flow) in the context of a mild CoA.

The paper is organized as follows: in section 2.1, the computational methods for the three different solvers utilized in this study are described including methods for data sampling and analysis. Section 2.2 presents a mesh sensitivity analysis conducted to ensure consistency and convergence of the CFD solution. In section 2.3, the method for implementing a 3-element windkessel model at the descending aorta (DAo) is described for the purpose of analyzing transient differences in pressure gradients across the CoA between the rigid wall models and one that considers compliance. Finally, section 3 presents the pressure gradients and velocity fields at rest and stress conditions, contrasting results from the three solvers along with a discussion of the limitations of this study in section 4.

2   Methods

2.1   Inter-solver Variability – A Verification Study

The numerical solution to the Navier-Stokes (NS) equations modeling incompressible, viscous flow may be arrived at in a discretized fluid domain by using multiple established numerical methods[10-13], within an FEM, FDM or FVM discretization framework. For this study, simulations were conducted in FLUENT 14.5 (ANSYS Inc., Canonsburg, PA, USA) - a FVM solver, COMSOL 4.2a (COMSOL Multiphysics Inc., Burlington, MA) – FEM solver and an in-house FDM cardiovascular flow solver implementing an unsteady artificial compressibility numerical method, each employing second-order spatio-temporal discretization schemes, under assumptions of incompressible, Newtonian fluid domain with rigid, impermeable walls. The aorta CoA model examined in this study was segmented from a contrast enhanced magnetic resonance angiography (MRA) and was provided as a STL file for the CFD Challenge. The STL surface mesh was converted into solid model constituted of NURBS patches after conducting minimal surface correction operations and finally exported as an IGS file for purposes of universal compatibility with different solvers. Direct numerical solutions (DNS) CFD was performed without considering a turbulence model, as per the specifications of the CFD challenge, for identical rigid-wall geometry, inlet flow rates (as provided for rest and stress conditions) with a plug profile and outlet mass-flow splits. Outlet mass-flow splits of 17%, 8%, 10% and 65% were applied for the rest condition and 25%, 5%, 11% and 59% were applied for the stress condition, at the Innominate, left carotid, left subclavian and DAo branches, respectively.

Finite Volume Solver: FLUENT
FLUENT is a FVM code. The IGS geometry was meshed in ANSYS Workbench 14. The curvature-based advanced mesh size function tool was used to obtain better mesh resolution in the region of the coarctation. A mesh containing ~500,000 tetrahedral elements was considered for the final FLUENT simulations reported in this study. A mesh sensitivity analysis was also performed as described in section 2.2. A User Defined Function (UDF) was implemented to input pulsatile mass flow rate waveform at the inlet for rest and stress cases. The pressure at the inlet was set to 63.35 mmHg for rest conditions and 64.3 mmHg for stress conditions. For all the FLUENT simulations, the pressure-velocity coupling algorithm was set as SIMPLE which implements a pressure based segregated algorithm. The SIMPLE algorithm uses a relationship between velocity and pressure corrections to enforce mass conservation and to obtain the pressure field. A second-order discretization of pressure and momentum terms was employed. The pressure discretization method was set as PRESTO!, recommended for complex geometries which induce swirl flow such as the aorta model. A second-order implicit time advancement with a fixed time step of 0.016 sec was chosen.

Finite Element Solver: COMSOL
COMSOL Multiphysics is a FEM code. A mesh containing ~600,000 and ~1,000,000 tetrahedral elements with an imposed boundary layer mesh was considered for the final rest and stress case COMSOL simulations respectively. The inlet flow waveform (15 Fourier-term waveform reconstructions) was specified at the inlet via COMSOL's live-link capabilities using MATLAB 2011b (The MathWorks Inc., Natrick, MA). A constant pressure was imposed at the inlet of the model as per the given pressure waveform. Simulations were conducted using the laminar flow module and a time dependent study. Furthermore, a preconditioned generalized minimal residual method (GMRES) iterative solver for the flow field at each time step and a constant (Newton) nonlinear solver time marching were implemented. Also, to aid in convergence for a highly nonlinear problem, the Jacobean matrix was updated after every iteration.

Finite Difference Solver- In-House Solver

DNS was also performed using a second-order accurate, in-house FDM, artificial compressibility numerical solver. This solver has been used extensively for image-based hemodynamic modeling and incorporates a validated multi-grid artificial compressibility numerical solver [8, 14, 15]. Flow was simulated on a high-resolution unstructured Cartesian immersed boundary grid composed of ~500,000 uniformly spaced nodes, with an average node spacing resolution of 0.02 mm, which was generated after immersing the surface model in a Cartesian grid of 498 x 126 x 156 cubical elements. Computations were performed using normalized spatial and temporal units. The temporal resolution was considered as 0.01 simulation time units i.e. ~O (10^{-4}) sec. A second order interpolation scheme was employed in order to obtain inlet conditions based on the input discrete cardiac cycle data. Mean pressure at the inlet was set to the mean of the given pressure waveforms.

Data Sampling

For the purpose of standardizing data collection a plane proximal to CoA and plane distal to CoA were defined as specified in Table 1. All results for pressure gradient across the CoA were reported at these planes and data was gathered for the 5th cardiac cycle, in order to ensure damping of initial transients.

Table 1. Planes before and after CoA defined by a point and through through the plane

	Origin	Normal to plane
Proximal plane	(188.96, 40.18, 253.22)	(0.98,-0.09,-0.19)
Distal plane	(261.97,23.56,277.10)	(0.99,-0.03,-0.14)

2.2   Mesh Sensitivity Analysis

A fine mesh (Fig 1a, 1c) consisting of ~500,000 elements and a coarse mesh (Fig 1b) consisting of ~250,000 elements were considered for the mesh sensitivity study in case of FLUENT. Flow was simulated for these meshes as described in section 2.1. The differences in pressure values for the two meshes for stress case are presented in Table 2 and Fig 2. Minor changes were observed before and after the CoA as a result of increasing the mesh size in FLUENT. As the mesh was refined similar consistency was observed in the reported pressure gradients for the in-house solver. In case of COMSOL, a finer mesh was required to obtain accurate results as per mesh sensitivity convergence tests. A mesh containing ~600,000 and ~1,000,000 tetrahedral elements was considered for the final rest and stress case COMSOL simulations respectively. Even for such fine meshes, the pressure results did were only approaching convergence but mesh refinement. Since the aim of this study was to compare pressure gradients using three solvers for similar mesh sizes, the mesh wasn't refined beyond this point. The FLUENT simulation results can be considered most accurate in this study since the pressure results converged appropriately as the final mesh density was approached.

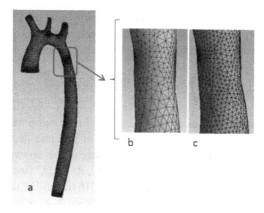

Fig. 1. a) Meshed aorta model used for the FLUENT simulation. Coarse (b), and fine (c) mesh seen at the coarctation.

Table 2. Reported area averaged pressure values at stress condition for mean flow proximal and distal to CoA for different number of mesh elements

Number of mesh elements	Area averaged pressure proximal to CoA (mmHg)	Area averaged pressure distal to CoA (mmHg)
250,000	61.075	48.275
500,000	61.15	46.83

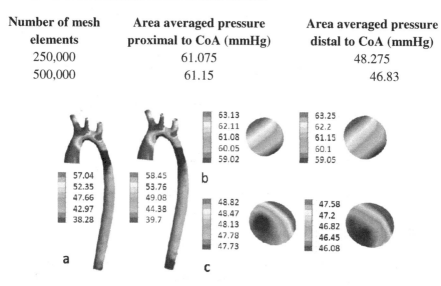

Fig. 2. Difference in pressure gradient for coarse (left) and fine (right) mesh at the walls (a), proximal plane (b) and distal plane (c), under rest-state conditions

2.3   3-Element Windkessel Modeling at DAo

The purpose of the windkessel modeling was to investigate the influence of compliance of walls on mean pressure gradients as well as its transient variation, only for the rest-state condition, in contrast with the rigid-wall CFD simulations described earlier (section 2.1) which did not factor compliance at all. In an effort to compensate for the compliant effect of the arterial walls and thereby tune the CFD results to match physiological pressure estimates [16, 17], a viscoelastic three element Windkessel model (Fig 3) was implemented based on the following equation, at the DAo:

$$\left(1 + \left(\frac{R_1}{R_2}\right)\right) Q_i + C\, R1\, \frac{dQ_i}{dt} = \frac{P_i}{R_2} + C\, \frac{dP_i}{dt} \tag{1}$$

where, R_1 is the primary resistance adjusted to yield required pressure at the baseline flow, C is compliance set such that time constant of pressure decay (R_1C) is around 2.5s. This represents the systemic arterial impedance at the downstream of the head-neck vessels and DAo. For this study, values of R_1, R_2 and C (see Fig 3) were considered as 1527mmHg-s/L, 100mmHg-s/L and 0.00164L/mmHg respectively. R_1 was tuned so as to match the given rest-state outlet mass flow splits without the compliance being factored in. Solving the linear algebraic-differential equation system on the model interface, the lumped parameter model represents the effect of the downstream domain as a semi-implicit function of pressure, which was used to couple the upstream three-dimensional numerical domain with the downstream analytic domain.

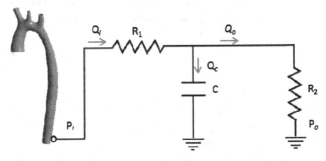

Fig. 3. Windkessel boundary condition imposed at the DAo

3   Results and Discussion

A solver-average mean pressure gradient of 2.79±0.7 mmHg was observed for the rest case as compared to 17.73±6.3 mmHg for the stress case as an average for all 3 rigid wall simulations (Table 3). An inter-solver variability of 16.9% was reported for mean pressure gradient for rest conditions and 23.7% was reported for mean pressure gradient for stress conditions. Exact flow splits were obtained at every time step for FLUENT simulations and in-house solver owing to the applied mass flow-split outflow boundary conditions. For COMSOL, computed flow splits were matched within 2.5% error for the rest case and within 1.2% for the stress. Mean flows of 10508, 4925, 6204 and 40635 mm³/s were measured for rest case and 55732, 11146, 24522 and 131527 mm³/s for measured for stress case through outlets one to four respectively. Renderings of the pressure field across the CoA from each of the three solvers (sections 2.1) can be seen in Fig 4. The mean pressure drop computed using Bernoulli principle across the coarctation using the mean CFD velocities observed in the plane distal to the CoA was ~18mmHg for rest conditions and ~90mmHg for stress conditions at peak flows which was an overestimation (~2 times higher) as compared to the pressure field reported from rigid wall CFD simulations.

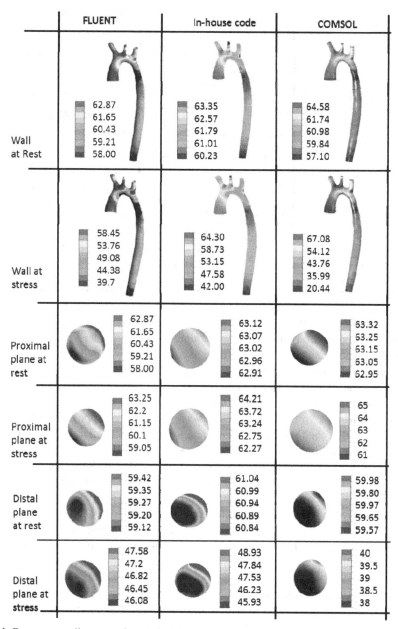

Fig. 4. Pressure gradients as observed at the wall, proximal and distal planes for FLUENT, in-house code and COMSOL

Table 3. Maximum, minimum and mean pressure at inlet and Dao including pressure drop (dP) across proximal and distal plane in mmHg

		Rest			Stress		
		FLUENT	In-house	COMSOL	FLUENT	In-house	COMSOL
Systole	Inlet	83.92	83.92	83.92	123.35	123.35	123.35
	Proximal (P)	81.46	83.06	82.91	111.16	118.50	116.03
	Distal (D)	68.13	66.92	68.33	70.52	51.84	26.33
	DAo	59.83	61.80	65.00	34.20	28.10	3.892
	dP = P – D	13.34	16.15	14.58	40.64	66.66	89.7
Diastole	Inlet	49.68	49.68	49.68	36.77	36.77	36.77
	Proximal	49.67	50.07	49.68	36.87	37.00	36.68
	Distal	49.48	49.99	49.68	39.07	37.36	36.63
	DAo	49.38	49.78	49.69	39.38	37.46	36.59
	dP = P – D	0.19	0.08	0.00	-2.20	-0.36	0.05
Mean flow	Inlet	63.35	63.35	63.35	64.30	64.30	64.30
	Proximal	62.96	63.02	63.12	61.34	63.24	62.34
	Distal	60.12	60.94	59.66	46.70	47.53	38.29
	DAo	58.55	60.23	58.64	35.52	42.01	32.22
	dP = P – D	2.84	2.08	3.46	14.64	15.71	24.04

In order to compare the nature of vortical structures in the DAo flow field between rest and stress conditions, helicity was computed as a flow derived parameter and rendered as isocontours (Fig 5a) for time-averaged inflow conditions. Helicity was computed as the normalized magnitude of the dot product between vorticity and velocity vectors at each node in the computed flow field. Positive helicity (highlighted red) indicates right handed vortical structures and negative helicity (highlighted blue) indicates left handed vortical structures. It was observed that opposing vortical structures are created as the flow enters the coarctation in a very similar formation for both rest and stress conditions in case of mild CoA. This was distinct from the opposing vortical structures that are formed which naturally spiral helically in the DAo by virtue of the curvature of the transverse aortic arch. However, the vortical structures distal to the modeled mild CoA were not as pronounced as seen for more severe CoA cases previously reported to demonstrate distinct downstream jet-flow effects[8].

Similar max-normalized cross-sectional velocity flow profiles were observed for the rest and stress conditions at the studied proximal and distal planes (Fig 5b) at rest and stress conditions. A shift in flow streams toward the outer curvature of the aortic arch was observed for stress conditions (viz. higher flow rates), leading to an elongated low-pressure pressure region distal to the CoA in contrast with the rest case (Fig 5b). Comparatively higher pressures were observed at the outer curvature of the aortic arch as seen on the proximal and distal plane (see Fig 4).

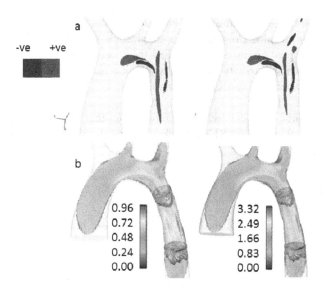

Fig. 5. Helicity isocontours (a) and velocity profiles (in m/s) (b) computed at mean pressures for rest (left) and stress (right) conditions

Pressure gradient across the CoA from the windkessel model showed transient variation at the peak systolic instant as an effect of factoring in compliance at the DAo as compared with the rigid wall simulations. Further, overall pressure values throughout the aorta model were found to increase significantly despite the mean pressure gradient not being affected considerably. This is congruent with the nature of the windkessel model for the cycle average pressure gradient calculation; as $dP_i/dt = 0$ and $dQ_i/dt = 0$ in equation (1) for the cycle average, which therefore reduces the equation to $P_i/Q_i = (R_1+R_2)$, therefore indicating that the pressure gradient should match that obtained for the cycle average obtained from rigid wall CFD simulations not factoring compliance. Fully coupled fluid structure interaction (FSI) methods can better predict clinically significant parameters such as pressure gradients for mean flows and wall shear stress for the arterial model as it considers the elastic properties of the arterial wall[18], while the windkessel approach still allows a tunable parameter to match physiological observations.

4   Conclusion

CFD is a powerful tool for simulation of altered hemodynamics in pathological anatomies and is investigated in this study as a diagnostic aid for evaluating pressure gradients across a patient-specific mild CoA. The downstream hemodynamics observed in the DAo was similar when simulated in three different rigid walls CFD solvers modeling viscous, incompressible blood flow, which in-turn were significantly different than pressure gradients computed from merely the inviscid Bernoulli formulation. However the rigid wall CFD approach is limited in the sense that regional vascular wall-compliance requires to be considered accurately in order to yield physiologically relevant results. Compliances applied at the outlets may not be enough for accurate pressure predictions and hence FSI must be adopted in future to take into consideration regional elasticity of arterial walls in a manner matched to wall motion observable from gated cine cardiac MRI studies.

Acknowledgements. We acknowledge funding from the Pittsburgh Supercomputing Center (research allocation BCS120006) for facilitating the parallel CFD simulations presented in this work.

References

1. Mullen, M.J.: Coarctation of the aorta in adults: do we need surgeons? Heart 89, 3–5 (2003)
2. Habib, W.K., Nanson, E.M.: The causes of hypertension in coarctation of the aorta. Annals of Surgery 168, 771–778 (1968)
3. Garne, E., Stoll, C., Clementi, M., Euroscan, G.: Evaluation of prenatal diagnosis of congenital heart diseases by ultrasound: experience from 20 European registries. Ultrasound in Obstetrics & Gynecology: The Official Journal of the International Society of Ultrasound in Obstetrics and Gynecology 17, 386–391 (2001)
4. Russo, V., Renzulli, M., La Palombara, C., Fattori, R.: Congenital diseases of the thoracic aorta. Role of MRI and MRA. European Radiology 16, 676–684 (2006)
5. Valverde, I., Staicu, C., Grotenhuis, H., Marzo, A., Rhode, K., Shi, Y., Brown, A.G., Tzifa, A., Hussain, T., Greil, G., Lawford, P., Razavi, R., Hose, R., Beerbaum, P.: Predicting hemodynamics in native and residual coarctation: preliminary results of a Rigid-Wall Computational-Fluid-Dynamics model (RW-CFD) validated against clinically invasive pressure measures at rest and during pharmacological stress. Poster Presentation at SCMR/Euro CMR Joint Scientific Sessions, February 3-6 (2011)
6. Varga, A., Kraft, G., Lakatos, F., Bigi, R., Paya, R., Picano, E.: Complications during pharmacological stress echocardiography: a video-case series. Cardiovascular Ultrasound 3, 25 (2005)
7. Mertes, H., Sawada, S.G., Ryan, T., Segar, D.S., Kovacs, R., Foltz, J., Feigenbaum, H.: Symptoms, adverse effects, and complications associated with dobutamine stress echocardiography. Experience in 1118 patients. Circulation 88, 15–19 (1993)
8. Menon, P.G., Pekkan, K., Madan, S.: Quantitative Hemodynamic Evaluation in Children with Coarctation of Aorta: Phase Contrast Cardiovascular MRI versus Computational Fluid Dynamics. In: Camara, O., Mansi, T., Pop, M., Rhode, K., Sermesant, M., Young, A. (eds.) STACOM 2012. LNCS, vol. 7746, pp. 9–16. Springer, Heidelberg (2013)
9. Eli Konen, N.M., Provost, Y., McLaughlin, P.R., Crossin, J., Paul, N.S.: Coarctation of the Aorta Before and After Correction: The Role of Cardiovascular MRI. American Journal of Roentgenology 182, 1333–1339 (2004)
10. Rannacher, R.: Finite Element Methods for the Incompressible Navier-S tokes Equations. Advances in Mathematical Fluid Mechanics, 191–293 (2000)
11. Chung, T.J.: Transitions and interactions of inviscid/viscous, compressible/incompressible and laminar/turbulent flows. International Journal for Numerical Methods in Fluids 31 (1999)

12. Wesseling, P., S.A., Vankan, J., Oosterlee, C. W., Kassels, C.G. M.: Finite discretization of the incompressible Navier-Stokes equations in general coordinates on staggered grids. Presented at the 4th International Symposium on Computational Fluid Dynamics, Davis, CA (September 1991)

13. Neal, T., Frink, S.Z.P.: Tetrahedral finite-volume solutions to the Navier-Stokes equations on complex configurations. International Journal for Numerical Methods in Fluids 31, 175-187

14. Menon, P.G., Yoshida, M., Pekkan, K.: Presurgical evaluation of fontan connection options for patients with apicocaval juxtaposition using computational fluid dynamics. Artificial organs 37, E1–E8 (2013)

15. Yoshida, M., Menon, P.G., Chrysostomou, C., Pekkan, K., Wearden, P.D., Oshima, Y., Okita, Y., Morell, V.O.: Total cavopulmonary connection in patients with apicocaval juxtaposition: optimal conduit route using preoperative angiogram and flow simulation. European Journal of Cardio-thoracic Surgery: Official Journal of the European Association for Cardio-thoracic Surgery 44, e46–e52 (2013)

16. Westerhof, N., Lankhaar, J.W., Westerhof, B.E.: The arterial Windkessel. Medical & Biological Engineering & Computing 47, 131–141 (2009)

17. Cappello, A., Gnudi, G., Lamberti, C.: Identification of the three-element windkessel model incorporating a pressure-dependent compliance. Annals of Biomedical Engineering 23, 164–177 (1995)

18. Reymond, P., Crosetto, P., Deparis, S., Quarteroni, A., Stergiopulos, N.: Physiological simulation of blood flow in the aorta: Comparison of hemodynamic indices as predicted by 3-D FSI, 3-D rigid wall and 1-D models. Medical Engineering & Physics 35, 784–791 (2013)

CFD Challenge: Predicting Patient-Specific Hemodynamics at Rest and Stress through an Aortic Coarctation

Christof Karmonik[1], Alistair Brown[2], Kristian Debus[2], Jean Bismuth[1], and Alain B. Lumsden[1]

[1] The Methodist Hospital, Houston TX, USA
[2] cd-adapco, Irvine CA, USA

Abstract. The here presented work is part of a CFD challenge investigating the potential for computational fluid dynamics (CFD) simulations to predicted pressures and flows in an aortic coarctation during stress when conditions for the rest case are known. In our approach, we choose to couple a three element Windkessel model to the outlet boundary conditions. Good reproducibility of flow and pressures for the rest case were achieved. In the stress case, where only the inflow boundary condition was changed, baseline pressure was too high, indicating that the total resistance in the Windkessel models may need to be reduced. This would correspond to dilating the blood vessels as might be the result of a pharmacological stress test. Future work is needed to develop an optimization strategy to tune the Windkessel data for matching the clinical results.

Keywords: computational fluid dynamics, aortic coarctation, Windkessel model.

1 Introduction

The work described in this manuscript is a contribution to a computational fluid dynamics (CFD) challenge investigating hemodynamics at rest and stress through an aortic coarctation [1]. As was stated on the website for the challenge (http://www.vascularmodel.org/miccai2013/), narrowing of the aorta (coarctation, CoA) accounts for approximately 10 % of congenital heart defects in the western world. A consequence of the reduction in luminal cross section is the existence of high pressure gradients which lead to an increase in the cardiac workload. In addition to the degree of luminal narrowing, also the flow rate will influence the value of the pressure gradient. Hemodynamics, in particular the 4D velocity field (3D plus time), can be assessed non-invasively by medical imaging techniques such as phase contrast magnetic resonance imaging [2]. Due to the reduction in the luminal diameter, flow changes during stress may result in a several-fold increase compared to rest. To replicate physiological stress condition during exercise, a pharmacological stress-test is sometimes performed, which are not ideal for the patient as often unwanted side-effects such as palpitations, chest pain, shortness of breath, headache nausea or fatigue may occur.

O. Camara et al. (Eds.): STACOM 2013, LNCS 8330, pp. 94–101, 2014.
© Springer-Verlag Berlin Heidelberg 2014

An alternative or complementary approach may be therefore to assess changes in the pressure gradient during stress using computational simulations utilizing patient-specific information (flow rates and geometries). In this approach, the simulation is first performed at rest to ensure that the corresponding (measured) pressure gradient is reproduced. Through adequate modification of the boundary conditions for stress conditions, the pressure gradient during stress is then estimated from these modified simulations. In this approach, it has not yet been established, how to 'adequately' modify the boundary conditions for the stress case. A wide variety of simulation techniques utilizing different concepts of boundary conditions exist, which have yet not thoroughly assessed towards their applicability for the here described medical problem. In our approach, we coupled an implicit unsteady model to a three element Windkessel model (RCR) for a more accurate treatment of the outlet boundary conditions.

2   Materials and Methods

2.1   Geometry

The geometry of the aorta was provided by the organizers of the CFD challenge in the form of a stereolithographic (STL) file containing a 2D surface mesh (138,532 faces and 69,268 vertices). This geometry was originally extracted from a 3D contrast-enhanced magnetic resonance angiography (MRA) dataset (contrast agent Dotarem). The geometry included aortic inflow (INLET 1), supra-aortic vessel outflow (OUTLET 2 - 4) and descending aorta outflow (OUTLET 1). This STL file was imported into STAR-CCM+ 8.04. The outlet lengths of the great arteries were extruded to stabilize the solution and ensure the boundary conditions (BCs) were not being applied too close to the bifurcation from the aorta. The definition of outlet 2 in the STL file supplied was not planar As such the boundary was modified to fix this issue before extruding.

2.2   Computational Mesh

The final computational mesh constructed based on the provided STL file contained 2, 946, 175 polyhedral elements with 5 prism layers in the near wall region. A local refinement was used in the region of the co-arctation to improve the resolution of the flow structures as the blood accelerates through the constriction (figure 1).

2.3   Solver and Physics

The simulation was run in STAR-CCM+ 8.04. An implicit unsteady model using a segregated approach was used to solve the time-accurate Navier-Stokes equations. A second order backward Euler scheme was used to advance through time. Second order upwind schemes were used for convection. Time-step size was 0.001 seconds, fluid density was 1000 kg/m^3 and fluid viscosity was 0.004 Pa.

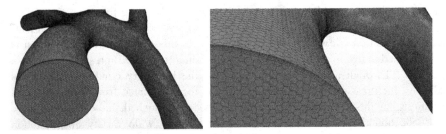

Fig. 1. Computational mesh (left: entire model, right: details showing prism boundary layer)

2.4   Boundary Conditions

Blood flow information was acquired using a phase contrast magnetic resonance imaging (2D, cardiac-gated, respiratory-compensated with through-plane velocity encoding). Cardiac output was 3.71 l/min, heart rate was 47 beats per minute with a duration of the cardiac cycle of 1.277 seconds. The organizers of the CFD challenge provided a 15-term Fourier reconstruction of the inflow waveform. Bases on these Fourier coefficients, (table 1) a flat velocity profile was defined at the inlet.

Table 1. Fourier coefficient for the inflow waveform (left: rest, right: stress)

	Rest	Stress
N	Q_n(x 10000)	Q_n(x 10000)
0	6.1811	4.0225
1	4.3744 - 9.2913i	1.0006 - 6.7808i
2	-4.4833 - 6.3505i	-4.4856 - 0.4473i
3	-5.5743 - 0.1296i	-1.1597 + 2.3282i
4	-2.1173 + 2.8983i	1.0939 + 0.9502i
5	-0.0324 + 1.0204i	0.2016 + 0.1259i
6	-0.7788 + 1.1164i	0.6237 + 0.1974i
7	0.4708 + 1.3418i	0.2460 - 0.6119i
8	0.7641 + 0.0220i	-0.2552 + 0.0533i
9	0.0140 + 0.1266i	-0.1241 - 0.0186i
10	0.1062 + 0.2667i	0.0092 + 0.0558i
11	0.3560 - 0.0207i	0.0262 + 0.1196i
12	0.1416 - 0.2389i	-0.0931 + 0.1268i
13	-0.2072 + 0.0149i	-0.1047 + 0.0311i
14	0.2269 + 0.1174i	-0.0932 + 0.0496i

A three element Windkessel model (ZCR) [3] was coupled to the outlet boundaries via a java macro (freely available on Macrohut http://macrohut.cd-adapco.com/phpBB3/viewtopic.php?f=4&t=321). The Windkessels were solved using a first order, backward Euler implementation and are coupled to STAR-CCM+ in an explicit manner. The Windkessel parameters are taken from the Brown et. al, where the values were tuned to match the clinical flow and pressure waves for this case (table 2).

Table 2. Windkessel parameters

Key	Outlet 1	Outlet 2	Outlet 3	Outlet 4	units
Z	1.31 e7	5.98 e7	1.59 e8	1.42 e7	$kg\ m^{-4}\ s^{-1}$
C	1.61 e-8	2.11 e-9	1.60 e-9	3.45 e-9	$m^{4}\ s^{2}\ kg^{-1}$
R	2.12 e8	1.47 e9	1.31 e9	5.87 e8	$kg\ m^{-4}\ s^{-1}$

3   Results

3.1   Rest Case

Good agreement between the clinically measured pressure at inlet and flow at outlet 1 (descending aorta) was obtained at end systole, peak flow and end diastole (figures 2).

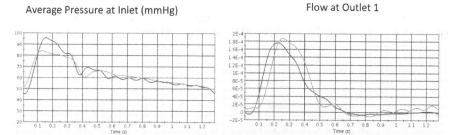

Fig. 2. Left: Average pressure at inlet (ascending aorta). Red: simulation results, green: clinically measured values provided by the organizers of the CFD challenge. **Right:**   flow at outlet 1 (descending aorta). Colors as in previous plot on top.

Wall shear stress values varied from maximum 110 Pa at peak flow to 1.8 Pa during end diastole (figure 3). Pressures varied from a maximum of 95 mmHg at peak flow to 50 mmHg at end diastole (figure 3).

3.2   Stress Case

For the stress case, inflow boundary conditions were adjusted as indicated by the data available at the CFD challenge website (table 2). The pressure obtained from the simulations at the ascending aorta and the flow rate at the descending aorta were overestimated (figure 4). Pressure values deviated by as much a factor of approximately 3 which was largely due to an overestimation of the baseline pressure (figure 4). Better agreement between the simulated and the clinically measured flow rates was achieved (figure 4).

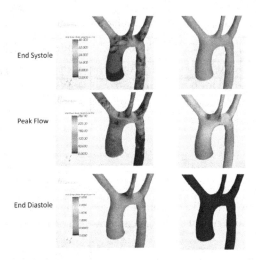

Fig. 3. Left: Wall shear stress at time points noted. **Right:** pressures. A large increase of pressure values proximal to the coarctation during peak flow can be appreciated.

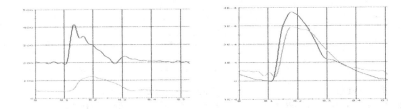

Fig. 4. Left: Average pressure at inlet (ascending aorta). Red: simulation results, green: clinically measured values provided by the organizers of the CFD challenge. **Right:** flow at outlet 1 (descending aorta). Colors as in previous plot on top.

4   Discussion

In this study, CFD simulations were performed within the framework of a CFD challenge. The model provided by the organizers of the study represented an aortic coarctation. For this condition, pressure gradients during stress conditions are difficult to obtain without avoiding side-effects for the patient, which justifies the use of computational simulations. Utilizing boundary conditions coupled to a Windkessel model, we were able to satisfactorily reproduce pressures at the inlet and flow waveforms at a the descending aorta (outlet 1) in the model. According to the instructions of the organizers of the challenge, a turbulent model was not used. However, it would not be non-physiological to expect some turbulent structures to occur as the flow accelerates through the coarctation and 'expands' out into the descending aorta. While this effect may occur during rest, it would even be more pronounced during stress.

Vessel wall was assumed rigid to be rigid which may be responsible for the over estimation of the pressure during systole. Including vessel compliance in the simulations would improve the correlation to clinical results. This effect may even be stronger in the Stress case where the compliance would have a larger damping effect on the peak pressure.

In general, CFD simulations of blood flow in vascular pathologies may be a viable alternative or can supply supplementary information to measurements provided the simulation results have been sufficiently validated and verified. Considerable work has been performed by many groups to qualitatively and quantitatively obtain insight in hemodynamics by simulations in a variety of patient-derived models: Pioneered early on [4-6], CFD simulations have recently been the focus of much attention in cerebral aneurysms [7-11] and abdominal aortic aneurysms [12-14] for assessing rupture risk. Usually, the workflow for such a simulation follows the following format: A suitable three-dimensional (3D) image set is obtained either from a medical image data using either computed tomographic angiography or MRI angiography. Both techniques display the aortic lumen hyperintense relative to the surrounding parenchyma. A surface model of the entire aorta is then created by applying image segmentation techniques. In addition to the technical challenge, this segmentation also includes judgment by the user which regions in the model are pertinent for obtaining accurate results and which can be safely omitted making this part of the workflow the most critical step. Commercial CFD software is capable of importing the surface model and from it, creating a computational mesh on which the governing partial differential equations, usually the Navier-Stokes equations, are solved. Recent postprocessing efforts have included converting the simulation results into image possibly to integrate the additional information from the simulations back into the clinical workflow [15]. Inflow and outflow boundary conditions are either taken from the literature or are individually measured in each patient.

Validation of the computational solutions is an essential step which should be performed for every different pathology. Without it, CFD will not gain broad acceptance in clinical research. Validation of computational results have been reported for cerebral aneurysms [16] [17] emphasizing the need for patient-derived geometries and inflow boundary conditions to arrive at realistic results. After validation has been performed, large-scale simulations may be performed towards their clinical efficacy for predicting disease progression and outcome. In this context, the CFD challenge posed here serves two purposes: First, it evaluates CFD for predicting clinical measurements (pressures and flow) based on given boundary conditions. Second, it also addresses variability of simulation results with different CFD software packages which is another important issue: Even is consistent and validated results are obtained with one solver code, there may still be variability when using another solver. This CFD challenge will give a good overview over the art of CFD for reproducing and predicting clinical measurements and others have to follow until a consensus on techniques and methodologies is achieved.

5   Conclusion

Our approach utilizing a three-element Windkessel model for simulating hemodynamics in the coarctation model of the CFD challenge resulted in good agreement between clinically measured pressures and flow at the descending aorta at rest but lacked agreement in the stress case. The reason for the discrepancies in the latter originated from suboptimal parameters of the Windkessel model not taking into account dilatation effects of the vessels. These effects are difficult to measure and consequently the adjustment of the boundary conditions is not trivial. Our findings show the potential but also current limitation of CFD. Future work is need to develop an approach where changes in boundary conditions caused by physiological effects can be predicated an integrated into the simulations.

References

[1] Shumacker Jr., H.B., Nahrwold, D.L., King, H., Waldhausen, J.A.: Coarctation of the aorta. Curr. Probl. Surg., 1–64 (February 1968)

[2] Frydrychowicz, A., Markl, M., Hirtler, D., Harloff, A., Schlensak, C., Geiger, J., Stiller, B., Arnold, R.: "Aortic hemodynamics in patients with and without repair of aortic coarctation: in vivo analysis by 4D flow-sensitive magnetic resonance imaging. Invest Radiol. 46, 317–325 (2011)

[3] Brown, A.G., Shi, Y., Marzo, A., Staicu, C., Valverde, I., Beerbaum, P., Lawford, P.V., Hose, D.R.: Accuracy vs. computational time: translating aor-tic simulations to the clinic. J. Biomech. 45, 516–523 (2012)

[4] Taylor, C.A., Hughes, T.J., Zarins, C.K.: Finite element modeling of three-dimensional pulsatile flow in the abdominal aorta: relevance to athero-sclerosis. Ann. Biomed. Eng. 26, 975–987 (1998)

[5] Taylor, C.A., Hughes, T.J., Zarins, C.K.: Effect of exercise on hemody-namic conditions in the abdominal aorta. J. Vasc. Surg. 29, 1077–1089 (1999)

[6] Taylor, T.W., Yamaguchi, T.: Flow patterns in three-dimensional left ventricular systolic and diastolic flows determined from computational fluid dynamics. Biorheology 32, 61–71 (1995)

[7] Castro, M., Putman, C., Radaelli, A., Frangi, A., Cebral, J.: Hemodynam-ics and rupture of terminal cerebral aneurysms. Acad. Radiol. 16, 1201–1207 (2009)

[8] Cebral, J.R., Hendrickson, S., Putman, C.M.: Hemodynamics in a lethal basilar artery aneurysm just before its rupture. AJNR Am J. Neuroradiol. 30, 95–98 (2009)

[9] Ford, M.D., Stuhne, G.R., Nikolov, H.N., Habets, D.F., Lownie, S.P., Holdsworth, D.W., Steinman, D.A.: Virtual angiography for visualization and validation of computational models of aneurysm hemodynamics. IEEE Trans. Med. Imaging 24, 1586–1592 (2005)

[10] Jou, L.D., Wong, G., Dispensa, B., Lawton, M.T., Higashida, R.T., Young, W.L., Saloner, D.: Correlation between lumenal geometry changes and hemodynamics in fusiform intracranial aneurysms. AJNR Am. J. Neuroradiol. 26, 2357–2363 (2005)

[11] Karmonik, C., Benndorf, G., Klucznik, R., Haykal, H., Strother, C.M.: Wall shear stress variations in basilar tip aneurysms investigated with computational fluid dynamics. In: Conf. Proc. IEEE Eng. Med. Biol. Soc., vol. 1, pp. 3214–3217 (2006)

[12] Kleinstreuer, C., Li, Z.: Analysis and computer program for rupture-risk prediction of abdominal aortic aneurysms. Biomed Eng. Online 5, 19 (2006)

[13] Kleinstreuer, C., Li, Z., Farber, M.A.: Fluid-structure interaction analyses of stented abdominal aortic aneurysms. Annu. Rev. Biomed. Eng. 9, 169–204 (2007)

[14] Li, Z., Kleinstreuer, C.: Blood flow and structure interactions in a stented abdominal aortic aneurysm model. Med. Eng. Phys. 27, 369–382 (2005)

[15] Karmonik, C., YJ., Spiegel, M., Redel, T., Mohammed, A., Horner, M., Kroeger, R., Grossman, R.G.: Effects of Inflow Variation in a Cerebral Aneu-rysm - An Image-based Approach for the Analysis of CFD Simulation Data. In: Asilomar Conference on Signals, Systems and Computers Asilomar, CA (2009)

[16] Karmonik, C., Klucznik, R., Benndorf, G.: Blood flow in cerebral aneu-rysms: comparison of phase contrast magnetic resonance and computational fluid dynamics–preliminary experience. Rofo 180, 209–215 (2008)

[17] Acevedo-Bolton, G., Jou, L.D., Dispensa, B.P., Lawton, M.T., Higa-shida, R.T., Martin, A.J., Young, W.L., Saloner, D.: Estimating the hemodynamic impact of interventional treatments of aneurysms: numerical simulation with experimental validation: technical case report. Neurosurgery, vol. 59, pp. E429-E430; author reply E429-E430 (August 2006)

A Multiscale Filtering-Based Parameter Estimation Method for Patient-Specific Coarctation Simulations in Rest and Exercise

Sanjay Pant, Benoit Fabrèges, Jean-Frédéric Gerbeau,
and Irene E. Vignon-Clementel

INRIA Paris-Rocquencourt, 78153 Le Chesnay, France and UPMC Université Paris 6,
Laboratoire Jacques-Louis Lions, 75005 Paris, France
Irene.Vignon-Clementel@inria.fr

Abstract. The 2nd CFD Challenge Predicting Patient-Specific Hemodynamics at Rest and Stress through an Aortic Coarctation provides
patient-specific flow and pressure data. In this work, a multiscale 0D-
3D strategy is tested to match the given data. The 3D outlet boundary
conditions for the supra-aortic vessels are represented by three-element
Windkessel models. In order to estimate the Windkessel parameters at
these outlets, a 0D lumped parameter model for the full aorta is considered. The parameters in such a 0D model are estimated by a sequential
estimation method, the unscented Kalman filter. The filtering approach
estimates the parameters such that the results of the 0D model closely
match the measured data: flow waveforms in the ascending and diaphragmatic aorta, mean flow rates in the supra-aortic vessels, and the pressure
waveform in the ascending aorta. Information from the 3D model is taken
into account in the full 0D model. This process is repeated for the two
separate cases of rest and stress conditions to estimate separate sets of
parameters for the two physiological states. Results such as the pressure
gradient across the coarctation, comparison with target values and more
detailed time or spatial variations are presented. Modelling choices and
assumptions about how the data are interpreted are then discussed.

1 Introduction

In the 'The 2nd CFD Challenge Predicting Patient-Specific Hemodynamics at
Rest and Stress through an Aortic Coarctation', the aim is to predict the pressure loss across a restrictive geometry, namely a coarctation of the aorta, at rest
and under stress, based on specific flow and pressure data measured for that
patient. This raises several modelling and numerical issues that have driven new
developments in the last decade. The geometry (assumed rigid) of this case consists of a patient aorta, with given inlet (aortic arch) and outlet (diaphragmatic
aorta) flow waveforms over one cardiac cycle, and three supra-aortic vessels for
which only the average flow is given. The inlet pressure waveform over one cycle is in addition given (systolic, diastolic and average pressures required to be
matched), and the challenge requires that the inlet flow waveform is imposed.

O. Camara et al. (Eds.): STACOM 2013, LNCS 8330, pp. 102–109, 2014.
© Springer-Verlag Berlin Heidelberg 2014

This set of data are incomplete to directly set the necessary boundary conditions for a 3D Navier-Stokes simulation. Hence, the first step is to complement this data by modelling assumptions. The easiest way from an implementation and numerical point of view is to assume a constant distribution of flow rates (matching the average given values) and velocity profiles, thus imposing velocities at each outlet nodes. However, these strong modelling assumptions may not generate the measured inlet pressure, and tuning of this distribution is then required. Besides, such a model has little predictive capabilities. In contrast, imposing a relationship between pressure and flow at the outlets such as the 0D Windkessel model, has been advocated, leading to multiscale modelling. It has the advantage of being more predictive (valid over a certain range of flow) and less constraining on the boundary velocities. At a given time point, the distribution of flows among the branches depends on the chosen 0D model (here Windkessel) parameters. Tuning of these parameters to match the given clinical data is, however, also necessary. Several strategies have been devised to achieve such goal. Manual tuning has been the first method used in practice, often requiring 0D experience [1]. This can become rapidly intractable, especially if there are many branches, and more precise data to match than average flow values. Automatic iterative approaches have thus been developed. In [2], a quasi-Newton method was used to tune the Windkessel parameters of the 3D model, to achieve a few pressure and flow waveform features. In [3], a fixed point algorithm was used to match given average regional flow rates (right and left lungs) and a measured pressure gradient (between the 3D inlet and the distal pressure of the outlet 0D models) on geometries involving dozens of branches. In [4], an adjoint-based method in a 3D bifurcating aneurysm lead to the estimation of Windkessel parameters to match systolic, diastolic and average pressure differences. In [6], wall displacement values were used to estimate the stiffness of the 3D fluid-solid interaction idealized aneurysm model and its outlet proximal resistance, based on a sequential estimation approach. Recently, [7] proposed two approaches, deterministic and Bayesian, for two inverse problems in hemodynamics: first, to identify arterial wall parameters by pressure measurements in a steady fluid structure interaction (FSI) problem; and second, for robust design optimisation of bypass grafts. Here we propose a different strategy to tune the 0D parameters of the 3D model, based on the filtering approach used in [6], but on a full representative 0D model. This has all the advantages of a sequential estimation approach, and in addition involves a fast-to-compute model that can be run over enough cycles to match given targets, while being enriched by a few 3D simulations.

2   Methods

2.1   Windkessel Parameter Estimation with a 0D Model

As will be seen in the next section, the outlet boundary conditions for the supra-aortic vessels in the 3D simulations are represented by three-element Windkessels. For Windkessel parameter estimation, a lumped parameter model of the

3D geometry is considered. Such a 0D model is shown in Figure 1: the 0D abstraction of the 3D domain is represented in black and the Windkesssel boundary conditions are represented in orange. Each segment of the aorta in the 3D domain is represented by a resistor and inductor, placed in series, to model the viscous and inertial effects, respectively. As will become clear in what follows, the authors believe that the FSI effects in the aorta play a significant role in determining the inlet pressure. Consequently, to model the FSI effects in a lumped manner, a capacitance (C_{fsi}) before the inlet is added.

In Figure 1 the variables $y_0 - y_{15}$ (in blue) and p_v (venous pressure) represent pressures, and the variables $y_{16} - y_{25}$ (in red) and y_c represent flow-rates. This electrical analog can be represented by a set of differential-algebraic equations, which can be solved in time, given the boundary conditions: q_{in}, q_{out}, and $p_v = 0$. Initially, the value of the 0D parameters to represent the 3D domain, i.e. the parameters R_{AA}, L_{AA}, R_{IN}, L_{IN}, R_{LC}, L_{LC}, R_{LS}, L_{LS}, R_{CoA}, L_{CoA}, R_{DA}, and L_{DA} are determined by the average geometric lengths and radii of the respective branches (c.f. [5] for the analytical expressions).

It should be noted that the authors did not initially include the inlet capacitance, C_{fsi}, but rather the results before its inclusion necessitated its presence. Without the inclusion of C_{fsi}, for the rest case, physiologically reasonable results were obtained in the the 3D simulations. However, the maximum errors between the 3D results and the quantities required to be matched were still around 20%. Moreover, for the stress case, the 0D model (without C_{fsi}) was unable to closely match the data provided, and the CFD simulations (with best found values of Windkessel parameters) showed abnormally high pressures in the ascending aorta. With these results the authors hypothesized that, perhaps, the reason for such an anomaly, particularly in the stress case, is due to the absence of FSI in the 3D model. This line of argument seems reasonable as it is required to match real measured data (with considerable movement of the arterial walls in general) with rigid wall assumptions in the CFD simulations. This presents a contradiction if FSI is non-negligible. Hence, to circumvent this difficulty, and yet not to break the rules of the challenge, the authors choose to model the FSI in a lumped manner by including a capacitance before the inlet of the 3D domain and keeping the 3D domain walls rigid as before. In what follows next, the parameter estimation methodology, which is primarily the same with or without C_{fsi}, except due to differences in the 0D model, is described.

A sequential estimation method, namely the unscented Kalman filter (UKF) [9–11], is employed for parameter estimation. In particular the library *Verdandi* [12] is used to implement the UKF. Owing to paucity of space, we refer for example to [6] for the details of such a method for parameter estimation in hemodynamic systems. The filtering approach estimates the parameters such that the results of the 0D model closely match the measured data in the patient. The pressure in the ascending aorta (y_0) and flow rates in the supra-aortic branches (y_{17}, y_{20}, and y_{23}) are prescribed as observations to the UKF (see [6] for the concept of state and observations) to estimate parameters. Since the UKF needs complete flow profiles for the supra-aortic branches, and only the mean

Table 1. Parameter estimates from the UKF procedure (all units in CGS)

Outlet boundary	Segment/ Parameter	Rest			Stress		
		R^p	C	R^d	R^p	C	R^d
	Innominate	459.1	1.87E-04	7364.4	193.3	6.39E-04	1069
	Left carotid	87.0	8.75E-05	14271.1	308.7	1.91E-04	5386.2
Dirichlet outlet	Left subclavian	584.4	1.05E-04	12628.4	512.2	2.72E-04	2498.9
	C_{fsi}		4.41E-04			2.31E-04	
	t_{shift}	0.083 s			0.002 s		
	Innominate	877.2	2.67E-04	7249.4	405.3	1.26E-03	852.3
	Left carotid	868.18	1.35E-04	14754.9	988.2	3.84E-04	4656.23
RCR outlet	Left subclavian	1443.7	1.59E-04	12858.5	885.1	7.03E-04	2015.7
	Descending aorta	177.3	5.9E-04	1851.1	70.9	2.44E-04	468.44
	C_{fsi}		6.2E-04			3.49E-04	
	t_{shift}	0.116 s			0.02 s		

flow rates are measured, the flow difference between the inlet and the outlet is split according to measured mean-flow splits at each time to generate time-dependent flow-curves in these branches. It is noted that the provided ascending aorta pressure leads (in time) the ascending aorta flow. In reality, this relationship is usually the opposite, i.e. the flow leads the pressure. Since the pressure and flow measurements were not taken simultaneously, it is imperative that the time-difference between the provided pressure and flow curves be accounted for. Hence, a parameter, t_{shift}, which reflects the time difference between measured flow and pressure curves is added to the estimation procedure. Once an estimate on the Windkessel parameters is obtained, a 3D CFD simulation (for details see next section) is run. From the results of this CFD simulation, the values of the 0D model parameters that represent the abstracted 3D domain are updated. In particular, a regression analysis is performed between the pressure drop in each segment versus the flow and flow-derivative through that segment to obtain revised estimates for the 0D resistance and inductance. With the updated 0D model, UKF estimation is performed again. This process is repeated until an acceptable match is found between the measured quantities and CFD results or the 0D parameters stop changing.

2.2   The 3-D Model

Blood is considered to be an incompressible Newtonian fluid modelled by the incompressible Navier-Stokes equations (solved using the finite element library *FELiSce*). The following boundary conditions are imposed:

Inlet boundary condition. As discussed in the previous section a lumped FSI capacitance is introduced before the inlet. The resulting pressure-flow relation, i.e. $P^{n+1} = P^n + \frac{dt}{C_{\text{fsi}}}\left(q_{\text{in}} - q^{n+1}\right)$, where q_{in} is the inlet flow, is imposed at the inlet.

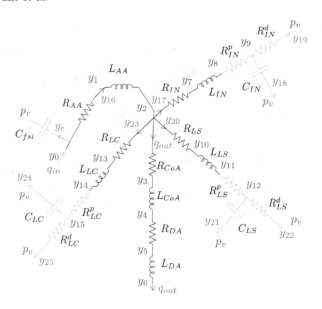

Fig. 1. The 0D model used: IN represents the innominate artery; LC represents the left carotid artery; LS represents the left subclavian artery; AA represents the ascending aorta; DA represents the descending aorta; CoA represents the aortic coarctation

The supra-aortic arteries. For the three upper arteries, explicit RCR Windkessel models are imposed. These boundary conditions are non homogeneous Neumann boundary conditions where the enforced pressure P is given by the following relations: $P^{n+1} = R_p q^n + P_c^{n+1}$, $P_c^{n+1} = P_c^n \left(1 - \frac{dt}{R_d C}\right) + q^n \frac{dt}{C}$.

Outlet boundary condition. For the descending aorta outlet, two boundary conditions are considered:

(a) *Dirichlet outlet*: In this case, an auxiliary steady Stokes equation, with natural boundary conditions at the inlet and outlets, is solved first. The resulting outlet velocity profile is scaled at each time to match the measured flow-rate, and is imposed at the outlet.

(b) *RCR outlet*: In this case an RCR Windkessel is imposed. It should be noted the Windkessel parameters are estimated as described in the previous section with the modified 0D model to include the Windkessel model at the outlet.

2.3   Mesh Generation

The 3-D mesh is generated from the surface mesh with the software *ghs3d*. The problems with the surface mesh are fixed, and subsequently a mesh-adapting code, *Feflo* [8], is run to both adapt and coarsen the mesh. The resulting 3-D mesh has 293016 tetrahedra. Feflo can adapt a 3-D mesh with respect to a velocity field. Mesh adaptation based on the velocity field in systole leads to a

375149 tetrahedra mesh for the rest case. For the stress case, a first 1138100 tetrahedra adapted mesh is obtained, followed by a second one with 291829 elements. Negligible differences between the solutions for these two meshes were found and thus computations were done on the coarser mesh.

3   Results and Discussion

As described in section 2.1 there are 11 and 14 parameters to be estimated for the two boundary conditions of 'Dirichlet outlet' and 'RCR outlet' at the descending aorta, respectively. The final estimated values of these parameters are tabulated in Table 1. For the two cases of rest and stress, three and seven iterations between the 0D and 3D models are needed, respectively, to achieve less than 10% errors between the 3D results and the measurements. Since a 0D model is used for parameter estimation, the time taken for one UKF run is negligible compared to the time taken for a full 3D simulation.

Figure 2 shows the pressures at the proximal and distal planes for both the two outlet boundary conditions. It is noticed that for the 'Dirichlet outlet' boundary condition, the pressure in the descending aorta shows oscillatory behaviour. This could be a result of the Dirichlet boundary condition being too constraining on the solution. The oscillations disappear when a Windkessel model is used at the outlet. For this latter boundary condition, the results of the 3D simulations and

Table 2. 3D results (for the 'RCR outlet' boundary condition) and comparison with targets: Flow rates are in L/min; Pressures are in mmHg; CoA represents the coarctation, f_s represents flow split as percentage of the inlet flow, P represents pressure, and ΔP represents pressure drop

Quantity	REST				STRESS			
	3D result	Target	Error	% err.	3D result	Target	Error	% error
Flow Innominate	0.624	0.624	0.000	0.00%	3.655	3.3550	0.300	8.94%
Flow Left-carotid	0.31	0.312	0.002	0.64%	0.748	0.6875	0.0605	8.80%
Flow Left-subclavian	0.358	0.364	0.006	1.65%	1.569	1.4575	0.1115	7.65%
Flow Desc. Aorta	2.417	2.410	0.007	0.29%	7.552	8.0300	0.4780	5.95%
f_s Innominate	16.8%	16.8%	0.0%	-	27.0%	24.8%	2.2%	-
f_s Left-carotid	8.4%	8.4%	0.0%	-	5.5%	5.1%	0.4%	-
f_s Left-subclavian	9.7%	9.8%	0.1%	-	11.6%	10.8%	0.8%	-
f_s Desc. Aorta	65.2%	65.0%	0.2%	-	55.8%	59.3%	3.5%	-
P_{max} Proximal plane	85.08	83.92	1.16	1.38%	112.58	123.35	10.77	8.73%
P_{min} Proximal plane	48.43	49.68	1.25	2.52%	34.81	36.77	1.96	5.33%
P_{mean} Proximal plane	63.84	63.35	0.49	0.77%	66.73	64.30	2.43	3.78%
P_{max} Distal plane	78.64	-	-	-	77.11	-	-	-
P_{min} Distal plane	48.43	-	-	-	30.35	-	-	-
P_{mean} Distal plane	61.84	-	-	-	54.08	-	-	-
ΔP_{max} CoA	11.8	-	-	-	41.34	-	-	-
ΔP_{mean} CoA	2.00	-	-	-	12.65	-	-	-

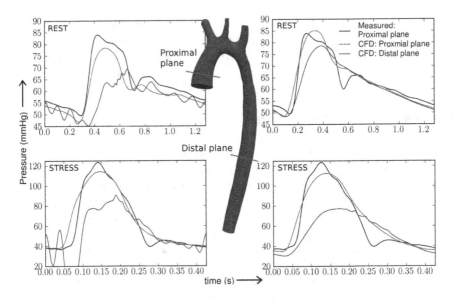

Fig. 2. Pressure results for the rest and stress cases; left: 'Dirichlet outlet' and right:'RCR outlet' boundary conditions

their comparison with the target (experimentally measured) values are shown in Table 2.

From Table 2 it can be noted that the errors between CFD results and experimentally measured quantities are less than 3% and 9% for the rest and stress cases, respectively. The higher errors for the stress case can be attributed to the fact that modelling of FSI in a lumped manner is relatively more difficult compared to the rest case as the flow gradients are significantly higher. It should also be noted that the current strategy for parameter estimation not only matches the systolic, diastolic, and mean pressures but also the full pressure profile of the measurements. Moreover, it is encouraging that for the rest case the estimated time shift, t_{shift}, corrects the possible anomaly of pressure leading the flow in the data provided for the challenge.

For the coarctation, mean and maximum pressure drops of 2.0 and 11.8 mmHg are reported for the rest case. Similarly, for the stress case, mean and maximum pressure drops of 12.65 and 41.34 mmHg are reported, respectively.

4   Conclusion

A complete framework, involving coupling of 0D and 3D models, for both parameter estimation and consequent 3D simulations is presented. The efficacy of such a method is demonstrated for the present case of patient-specific coarctation. The results for both the physiological states of rest and stress are shown to be in good agreement with the measured data: less than 3% and 9% errors for all measured quantities in the rest and stress cases, respectively. It is also found that a Dirichlet boundary condition at the outlet is significantly constraining on the solution and can result in pressure oscillations. In this sense, the use of Windkessel boundary conditions is reported to yield significantly better results. The proposed strategy of parameter estimation is also shown to be effective in dealing with some uncertainty in measurements, for example the time-difference between pressure and flow curves, if they are not measured simultaneously.

References

1. Pennati, G., Corsini, C., Cosentino, D., Hsia, T.-Y., Luisi, V., Dubini, G., Migliavacca, F.: Boundary conditions of patient-specific fluid dynamics modelling of cavopulmonary connections: possible adaptation of pulmonary resistances results in a critical issue for a virtual surgical planning. J. R. Soc. Interface 1(3), 297–307 (2011)
2. Spilker, R.L., Taylor, C.A.: Tuning multidomain hemodynamic simulations to match physiological measurements. Ann. Biomed. Eng. 38(8), 26351748 (2010)
3. Troianowski, G., Taylor, C.A., Feinstein, J., Vignon-Clementel, I.E.: Three-dimensional simulations in glenn patients: clinically based boundary conditions, hemodynamic results and sensitivity to input data. J. of Biomech. Engrg. 133(11), 111006 (2011)
4. Ismail, M., Wall, W.A., Gee, M.G.: Adoint-Based Inverse Analysis of Windkessel Parameters for Patient-Specific Vascular Models. J. of Comp. Physics 244, 113170 (2013)
5. Milišić, V., Quarteroni, A.: Analysis of lumped parameter models for blood flow simulations and their relation with 1D models. ESAIM-Math. Model. Num. 38, 613–632 (2004)
6. Bertoglio, C., Moireau, P., Gerbeau, J.-F.: Sequential parameter estimation for fluid–structure problems: Application to hemodynamics. Int. J. Num. Methods in Biomed. Engrg. 28(4), 434–455 (2012)
7. Lassila, T., Manzoni, A., Quarteroni, A., Gianluigi, G.: A reduced computational and geometrical framework for inverse problems in hemodynamics. Int. J. Num. Methods in Biomed. Engrg. 29(7), 2040–7947 (2013)
8. Loseille, A., Löhner, R.: Adaptive anisotropic simulations in Aerodynamics. In: 48th AIAA Aerospace Sciences Meeting, AIAA 2010-169, Olando, FL, USA (2010), http://www-roc.inria.fr/gamma/Adrien.Loseille/index.php?page=softwares
9. Julier, S., Uhlmann, J., Durrant-Whyte, H.F.: A new approach for filtering non-linear systems. In: American Control Conference, pp. 1628–1632 (1995)
10. Pham, D.T., Verron, J., Roubeaud, M.C.: A singular evolutive extended Kalman filter for data assimilation in oceanography. J. Marine Syst. 16(3-4), 323–340 (1998)
11. Moireau, P., Chapelle, D.: Reduced-order Unscented Kalman Filtering with application to parameter identification in large-dimensional systems. ESAIM Contr. Optim. Ca. 17, 380–405 (2011)
12. Chapelle, D., Fragu, M., Mallet, V., Moireau, P.: Fundamental principles of data assimilation underlying the Verdandi library: applications to biophysical model personalization within euHeart. Med. Biol. Eng. Comput. 1 (2013), http://verdandi.sourceforge.net/

A Finite Element CFD Simulation for Predicting Patient-Specific Hemodynamics of an Aortic Coarctation

Idit Avrahami

Ariel Biomechanics Center, Ariel University, Israel
iditav@ariel.ac.il

Abstract. This paper presents a numerical simulation of the flow characteristics through a patient-specific model of an aortic coarctation. The purpose of the study was to predict the pressure gradient at rest and at exercise conditions. The commercial package ADINA was used to numerically solve the governing equations using finite-elements methods. The model was based on the patient's MR angiography data. The boundary conditions imposed in the model included the flow and pressure waveforms acquired at the ascending aorta inlet and flow at the descending aorta outlet. Imposed flow waveforms at the supra-aortic vessels were estimated from the time-dependent difference between the ascending and descending aorta waveforms, and the flow distribution was dictated according to the total flow rates at each branch. The simulations considered two cases of rest and stress flow conditions. The time-dependent pressure in the proximal and distal planes and pressure gradients along the aorta are reported for rest and stress conditions.

1   Introduction

Coarctation of the aorta (CoA) is a constriction of the aorta. It is one of the most common diagnoses in congenital cardiac defects with prevalence variation from 5% to 11% of all congenital heart defects. It is characterized by significant systolic pressure gradient (> 20 mm Hg), resulting in an increased cardiac workload[1]. Pressure gradients increase mainly during stress conditions. CoA was traditionally treated by surgery, however it is increasingly replaced by catheter techniques such as balloon angioplasty in children and stents in adolescents and adults[2, 3].

Patient specific numerical simulations can assist in diagnosis and management of CoA by predicting the flow and pressure dynamics and the effect of the specific pathological anatomy on the patient's blood pressure [3, 4]. However, different numerical approaches might lead to different pressure predictions.

Therefore, in the scope of the MICCAI 2013 CFD challenge, multiple groups investigate the same case in order to analyze and compare different numerical approaches and to find the effect of method concept on the calculated patient-specific hemodynamics. The defined model is of a mild thoracic aortic coarctation of a 17 year old male. The purpose of the challenge is to compare different numerical approaches for the simulation of blood pressure gradients along a given aorta model and prescribed physiological conditions measured at rest and under exercise. In the

O. Camara et al. (Eds.): STACOM 2013, LNCS 8330, pp. 110–117, 2014.
© Springer-Verlag Berlin Heidelberg 2014

present manuscript, a computational fluid dynamics (CFD) simulation of the CoA hemodynamics is presented using the finite-element commercial software ADINA (ADINA R&D, Inc., MA).

2   The Numerical Model

The 3D patient-specific geometry (shown in Figure 1a) was based on data acquired using Gadolinium-enhanced MR angiography (MRA). The data segmentation was provided to the challenge in STereo Lithography (STL) file format, describing only the surface geometry of the 3D geometry.

The boundary representation file was imported into ADINA-M using the ADINA discrete boundary representation (BREP) feature and the body was adapted to the specified mesh. The resulted geometric mesh included the lumen of the ascending aorta (AscAo), the aortic arch, the coarctation, the descending aorta (DAo), and the three large branches: the innominate artery (Inn), the left common carotid (LCC) artery, and the left subclavian (LS) artery (see Figure 1b). The average diameter of each inlet and outlet in the model are listed in Table 1.

The boundary conditions included time-dependent proximal flowrate and pressure and distal flowrate waveforms, for rest and stress conditions. In addition, flow division between the three large branches was provided for the two cases, as listed in Table 1. From that data, time-dependent velocities were extracted for each outlet (see Figure 2). Velocity boundary conditions were imposed at the AscAo inlet and at the DAo, LCC and LS outlets.

Since the model assumes rigid walls, and the objective of the challenge is the pressure gradient, no pressure boundary conditions were specified. Time dependent pressure and flowrate values were extracted at the specified proximal and distal planes shown in Figure 3.

Each model was composed of 3,920,681 tetrahedral elements with refinements near the boundaries and at the supra-aortic vessels (see Figure 1b and c). For both cases, time steps of 0.01 sec were used to solve the time-dependent flow field along the cardiac cycle.

The flow and pressure fields in the lumen were calculated by numerically solving the equations governing momentum and continuity in the fluid domain:

$$\nabla \cdot \mathbf{V} = 0$$

$$\rho \frac{D\mathbf{V}}{Dt} = -\nabla p + \mu \nabla^2 \mathbf{V} + \rho \mathbf{g}$$

(1)

where p is static pressure, t is time, $\mathbf{V}$ is velocity vector, ρ and μ are density and the dynamic viscosity of blood, and $\mathbf{g}$ is vector of gravity. Blood was assumed homogenous, incompressible (with density $\rho = 0.001 \text{g/mm}^3$), and Newtonian (with viscosity $\mu = 0.004$ g/mm/s). The flow was assumed laminar. No gravity was employed.

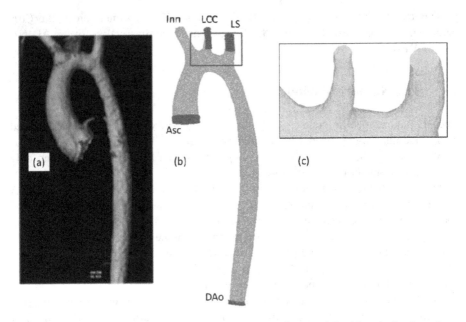

Fig. 1. The 3D patient-specific geometry (a), the numerical model with velocity boundary conditions (b) and a magnified view on the mesh near the upper vessels (c)

Table 1. Model data - Geometry dimensions and flowrate

	AscAo	Innominate	LCC	LS	DiaphAo
d [mm]	21.39	7.81	3.61	5.98	12.84
Total Flow at REST [L/min]	3.71	0.624	0.312	0.364	2.41
% flow at REST	100%	17%	8%	10%	65%
Total Flow at STRESS [L/min]	13.53	3.355	0.6875	1.4575	8.03
% flow at STRESS	100%	25%	5%	11%	59%

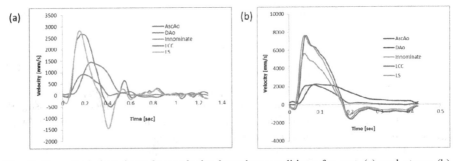

Fig. 2. Prescribed time-dependent velocity boundary-conditions for rest (a) and stress (b) conditions

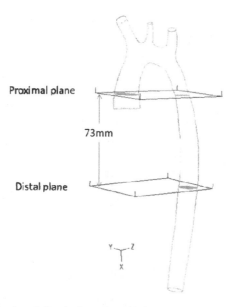

Fig. 3. Specified proximal and distal planes at which the pressure and flowrate results were acquired

For the simulations, a Dell PowerEdge R710 server was used with 16 cores. The computational time for the REST and STRESS simulations were 22 and 18 hours, respectively.

3   Results

Examples of the calculated pressure gradient for both cases are shown for systole and diastole in Figure 4. Average pressure values are listed as a function of time at the distal and proximal planes and the time-dependent gradient was calculated as the difference between them (see Figure 5). Mean and maximum pressure gradients and average flowrate are detailed in Table 2 .

In addition, time dependent flowrate at the proximal and distal planes were listed and compared with the specified measured flowrate waveforms (Figure 6). The resulted average flowrate was compared with the prescribed total flowrate ($Q_{prescribed}$) listed in Table 1 for both cases. The error ($\%err$) was calculated using:

$$\%err = \frac{Q_{prescribed} - \dfrac{\int Q_{calculated}(t)dt}{T}}{Q_{prescribed}} * 100\% \tag{1}$$

where $Q_{calculated}(t)$ is the flowrate at each time step and T is the time period. The average flowrate and errors are listed in Table 3.

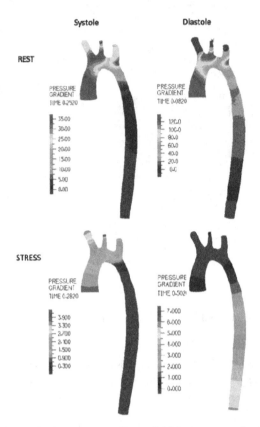

Fig. 4. Examples of the resulted pressure gradients field between proximal and distal planes under rest (top) and stress (bottom) conditions for Systole (left) and Diastole (right)

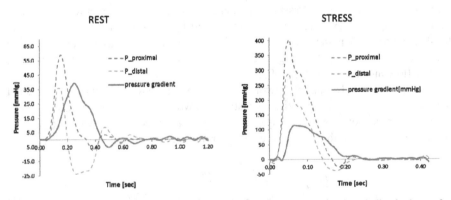

Fig. 5. Time-dependent pressure gradient as calculated from proximal and distal planes for REST (left) and STRESS (right) cases

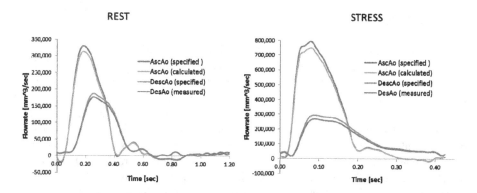

Fig. 6. Time dependent flowrate at the proximal and distal planes as specified for boundary conditions, and as calculated from the simulation results, for REST (left) and STRESS (right) cases

Table 2. Mean and maximum pressure gradients as calculated in the present study, compared with example results by LaDisa et al.

		REST	STRESS
Present study	Mean pressure [mmHg]	5.7	27.9
	Maximal pressure [mmHg]	28.1	112.5
LaDisa et al., 2011 [3]	Mean pressure [mmHg]	7	31
	Maximal pressure [mmHg]	22	73

Table 3. Mean flowrate for the two cases, compared with the prescribed total flow

	REST		STRESS	
	AscAo	DAo	AscAo	DAo
Mean flowrate [mm^3/s]	36870	56642	123770	217294
Calculated flowrate [mL/min]	2.21	3.40	7.43	13.04
Prescribed total flow [mL/min]	2.41	3.71	8.03	13.53
% error	8.2%	8.4%	7.5%	3.6%

4   Discussion

The results from the 3D simulations for rest and stress conditions reveal the high blood pressure gradient in the thoracic aorta due to the coarctation. In both cases there is a significant pressure gradient across the coarctation at peak systole of 39 mmHg under rest and 112.5 mmHg under stress conditions, respectively. The corresponding results for mean pressure gradient are 5.7 mmHg and 27.9 mmHg for rest and stress cases, respectively. In previously reported studies of moderate CoA by LaDisa et al.[3], maximal values of pressure gradients reached 22 mmHg and 73 mmHg at rest and exercise conditions, respectively. The mean values reported are 7 mmHg and 31 mmHg, which are a little higher than the results of the present study.

The extremely high values at peak systole of the exercise case are probably an overestimation of the real values. In order to make the simulation effort simpler and to reduce the variability in the result obtained by different groups, some major simplifications were defined for the challenge. Some of the more important simplifications are: the arterial wall was assumed to be rigid, blood was assumed Newtonian and the flow is assumed laminar.

The assumption of Newtonian fluid should be reasonable in this case, because the shear rates in the aorta are generally greater than $100 \ s^{-1}$ [5]. However, the effect of wall motion can be critical. The effect of wall motion was discussed by previous studies [6] and it was suggested that the flow in the aorta is a result not only of aorta geometric curvatures, but also of the motion of the aorta resulting from its attachment to the beating heart. In addition, simulations of aortic hemodynamics with fluid-structure interaction (FSI) approaches were found to predict better the pressure and wall shear stress in the aorta [7]. However, since the mechanical properties and thickness of the aortic wall are in doubt, simulations that include the passive wall motion due to its compliance are a great challenge. Therefore, this simplification was necessary to avoid a large difference between groups in the challenge.

The assumption of laminar flow is even more problematic. In an aortic coarctation, the blood flow can undergo a transition from a well-structured laminar state to a chaotic turbulent state, as the narrowing causes the velocity to increase[8, 9]. This transition is a challenge to model. Recent investigations partially succeeded to develop and use low-Reynolds turbulence models that showed good agreements between measurements and numerical results within accuracy of up to 10%.[9]. However these models required high computational resources (the simulations required 12 cardiac cycles and mesh sizes of 7 Million elements in order to ensure convergence).

Since the purpose of the challenge was to compare between different numerical approaches, the results should be compared to the physiological data in respect to the above simplifications.

References

1. O'Rourke, M.F., Cartmill, T.B.: Influence of aortic coarctation on pulsatile hemodynamics in the proximal aorta. Circulation 44(2), 281–292 (1971)
2. Rao, P.S.: Coarctation of the aorta. Current Cardiology Reports 7, 425–434 (2005)

3. LaDisa Jr., J.F., et al.: Computational simulations for aortic coarctation: representative results from a sampling of patients. Journal of Biomechanical Engineering 133(9), 091008–091008 (2011)
4. Ralovich, K., et al.: Hemodynamic assessment of pre- and post-operative aortic coarctation from MRI. In: Ayache, N., Delingette, H., Golland, P., Mori, K. (eds.) MICCAI 2012, Part II. LNCS, vol. 7511, pp. 486–493. Springer, Heidelberg (2012)
5. Fung, Y.C.: Biomechanics: mechanical properties of living. Springer (1993)
6. Jin, S., Oshinski, J., Giddens, D.P.: Effects of wall motion and compliance on flow patterns in the ascending aorta. Journal of Biomechanical Engineering 125(3), 347–354 (2003)
7. Crosetto, P., et al.: Fluid–structure interaction simulation of aortic blood flow. Computers & Fluids 43(1), 46–57 (2011)
8. Lantz, J., et al.: Validation of turbulent kinetic energy in an aortic coarctation before and after intervention-MRI vs. CFD. Journal of Cardiovascular Magnetic Resonance 15(suppl. 1), E46 (2013)
9. Lantz, J., et al.: Numerical and experimental assessment of turbulent kinetic energy in an aortic coarctation. Journal of Biomechanics (2013)

Traditional CFD Boundary Conditions Applied to Blood Analog Flow through a Patient-Specific Aortic Coarctation

Xiao Wang[1], D. Keith Walters[1,2], Greg W. Burgreen[1],
and David S. Thompson[1,3]

[1] Center for Advanced Vechicular Systems
[2] Department of Mechanical Engineering
[3] Department of Aerospace Engineering
Mississippi State University, Starkville MS 39759, USA

Abstract. Flow of a blood analog is modeled through a patient-specific
aortic coarctation using ANSYS Fluent software. Details of the patient
data (aortic geometry and prescribed flow conditions) were provided by
the MICCAI-STACOM CFD Challenge website. The objective is to predict a blood pressure difference across the rigid coarctation under both
rest and exercise (stress) conditions. The supplied STL geometry was
used to create coarse and fine viscous meshes of 250K and 4.6M cells. Our
CFD method employed laminar, Newtonian flow with a total pressure
inlet condition and special outlet BCs derived from reconstructed flow
waveforms. Analysis setup and outlet BCs were treated as a traditional
non-physiological CFD problem. CFD results demonstrate that the supplied AscAo pressure waveform and flow distributions are well matched
by our simulations. A non-uniform pressure gradient field is predicted
across the coarctation with strong interactions with each supra-aortic
vessel branch.

Keywords: computational fluid dynamics, aortic arch flow.

1 Introduction

Physiology of an aorta involves a significant compliant volume change driven by
pressure/flow pulses generated by a beating left ventricle. The aorta acts as the
central pressurized arterial blood plenum that supplies all major organ systems in
the human circulation. CFD analysis of an isolated aorta is challenging because
both its dynamic shape and its interaction with the circulation must be realistically approximated. This study performs a highly simplified but physically valid
fluid dynamic simulation of an isolated rigid aorta. Directly using the BC data
supplied by the website (namely, time-varying pressure/flow profiles at the AscAo and flow profiles at the DescAo), a pressure field is predicted that identically
matches the AscAo and DescAo profiles subject to the basic constraint that the
three supra-aortic vessels account for the flow differences required to conserve
mass. This study intentionally neglects key features defining a physiologically

O. Camara et al. (Eds.): STACOM 2013, LNCS 8330, pp. 118–125, 2014.

realistic aorta, particularly, its time varying shape compliance and interaction with fluid dynamically nonlinear vascular subsystems of the circulation. The objective is to generate baseline flow solutions of a simplified idealized aorta for comparisons to physiological data and to other CFD solutions that involve more complex modeling approaches.

2   CFD Model Setup

The commercial software package ANSYS Fluent v14.0 was used to perform the simulations. Laminar flow was assumed for the flow throughout the entire cardiac cycle. The working fluid was a Newtonian blood analog with constant density of 1000 kg/m^3 and viscosity of 0.004 kg/ms. The simulations used second-order spatial discretization for the momentum equations, the second order interpolation scheme for discretization of the pressure, and the SIMPLE scheme for pressure-velocity coupling. Unsteady terms were discretized using a second-order implicit scheme.

The time-accurate simulations were based on a repeating cardiac cycle using 400 uniform time steps per cycle. To ensure accuracy of the unsteady solution, 100 subiterations were used at each time step. For each case, unsteady calculations were performed for a time period of two cardiac cycles to ensure that a periodic flow pattern was achieved. Solutions extracted from the second cardiac cycle are reported and analyzed.

At the inlet and outlet boundary surfaces, unsteady flow conditions were incorporated into the FLUENT solver using user-defined function subroutines. The prescribed ascending aortic pressure variation was applied via a spatially uniform, time-varying total pressure boundary condition derived from the supplied waveforms of the ascending aortic pressure and flow rate. A temporally-varying, spatially-uniform velocity boundary condition based on the provided waveform was prescribed at the descending aortic outlet. Flow velocities were also specified at the three supra-aortic vessels (Innominate, LCC, LS). These velocities were based on the instantaneous net flow rate, i.e., the difference between the flow rate entering the domain and the flow rate exiting the domain at the descending aortic outlet. The net flow rate was partitioned among the three outlets to preserve the ratios provided in Table 1. The ratio of flow through each supra-aortic vessel was assumed to be a constant fraction of the total flow through all three vessels. This approach ensures that the correct average flow rate is obtained at each inlet/exit. No-slip conditions were applied at wall boundaries. All walls were considered to be strictly rigid with no fluid-structure interaction.

Volumetric flow rates and spatially-averaged static pressures at each boundary were recorded at every time step to characterize the unsteady flow. In order to evaluate the coarctation pressure gradients consistently with the invasive pressure wire measurements, pressures in the proximal and distal planes at the required locations were recorded at each time step as well.

Table 1. CFD-predicted total flow rates (L/min) through outlets

Outlet	Rest Condition			Stress Condition		
	Supplied	CFD		Supplied	CFD	
		Fine Mesh	Coarse Mesh		Fine Mesh	Coarse Mesh
AscAo	3.71	3.73	3.70	15.53	13.64	13.54
Innominate	0.624	0.630	0.625	3.355	3.384	3.358
LCC	0.312	0.306	0.294	0.6875	0.6989	0.6716
LS	0.364	0.372	0.368	1.4575	1.4950	1.4775
DiaphAo	2.41	2.42	2.41	8.03	8.06	8.04

3   Mesh Refinement

A mesh refinement study was performed by comparing results computed on a coarse mesh (250k cells) and a fine mesh (4.57M cells). Both meshes were tetrahedral-dominant with five layers of prismatic boundary layer cells. Simulations were conducted using both meshes under the stress condition. Comparing the time-resolved spatial-average pressure at each boundary surface, slight differences are observed only in the descending pressure data at the descending outlet (Fig. 1(a)). Based on these results the fine mesh was judged to be sufficient for analysis of results.

4   CFD Results

The unsteady simulations were conducted for two physiologic states of a patient, in rest condition and in stress condition. The heart rate is 47 beats per minute under rest condition, and 141 beats per minute under stress condition. The corresponding cardiac cycle is 1.277 seconds for rest condition, and 0.45 seconds for stress condition.

4.1   Rest Condition

For the rest condition, unsteady simulations were performed for a time period of two cardiac cycles (cardiac cycle 1.277 sec, total flow time 2.554 sec) using total 800 time steps with a uniform step size of 3.1925e-3 sec.

Fig. 1(b) shows that the static pressure obtained at the ascending boundary in the CFD simulations and the supplied ascending aortic pressure are in good agreement, which indicates that the total pressure specified at the inlet face matches the given inlet condition. The unsteady flow rate at the ascending inlet should automatically match the supplied ascending aortic flow waveforms when all the outlet boundary flow rates are specified based on the given flow conditions. Fig. 1(c) depicts the time history of predicated pressure gradients responding to the flow waveforms. Solutions of pressure and velocity fields were extracted at the instant of peak flow (0.19 sec) as shown in Figs. 2, 3 and 4.

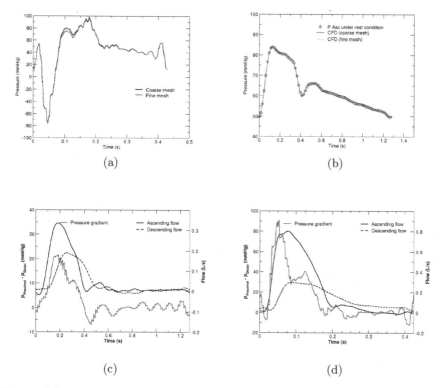

Fig. 1. (a) Static pressure at descending outlet shows only slight differences between coarse and fine mesh. (b) Static pressure at ascending inlet under rest condition. (c) Pressure gradient and flow waveforms under rest condition. (d) Pressure gradient and flow waveforms under stress condition.

The aorta wall pressure shown in (Fig. 2) indicates that the pressure distribution in the supra-aortic vessels are related to their size. The left carotid artery (LCC) is the narrowest branch with lower flow rate and lower pressure. A cutting plane shown in Fig. 3 details the pressure change in the aorta interior. The higher ascending pressure pumps blood into the supra-aortic vessels. As the flow of blood is turning at the arch of aorta, pressure drops quickly until it enters the descending aorta, thereafter flow pressure is gradually decreasing to reach the lowest value at the descending outlet.

Contours of flow velocity magnitude in Fig. 4 reveal two regions of flow separation. The arch of the aorta induced large flow separations due to the sudden change of flow direction. Another thin layer of separation is observed on the wall of the descending aorta. The coarctation (i.e. narrowing or pinching) in the aorta of this patient, between the aortic arch and descending aorta, forms a shape similar to a converging-diverging nozzle. The diverging wall produces adverse pressure gradients on the boundary layer, and encourages flow separation, which helps to explain the separations that occur on the diverging section of the aorta wall.

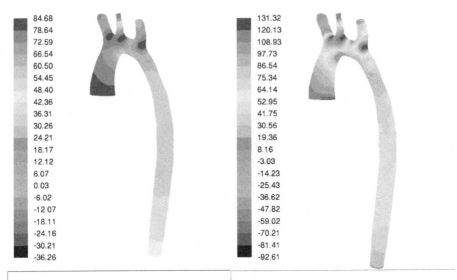

Contours of Static Pressure (mmhg) (Time=1.9155e-01) Contours of Static Pressure (mmhg) (Time=7.9688e-02)

Fig. 2. Aorta wall pressure at the instant of peak ascending flow at rest (left) and stress (right) condition

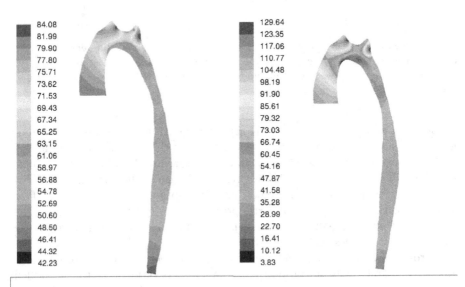

Contours of Static Pressure (mmhg) (Time=1.9155e-01) Contours of Static Pressure (mmhg) (Time=7.9688e-02)

Fig. 3. Pressure in aorta cutting plane at the instant of peak ascending flow at rest (left) and stress (right) condition

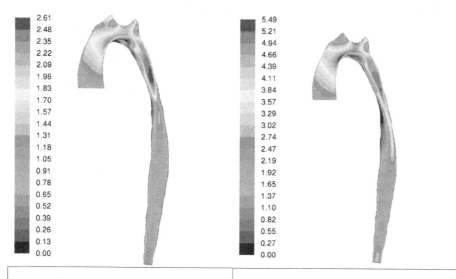

Contours of Velocity Magnitude (m/s) (Time=1.9155e-01)   Contours of Velocity Magnitude (m/s) (Time=7.9688e-02)

Fig. 4. Velocity magnitude in aorta cutting plane at the instant of peak ascending flow at rest (left) and stress (right) condition

4.2   Stress Condition

For the stress condition, unsteady simulations were also performed for a time period of two cardiac cycles (cardiac cycle 0.425 sec, total flow time 0.85 sec) using a total of 800 time steps with a uniform time step of 1.0625e-3 sec. The predicted pressure gradient corresponding to the supplied flow rate is presented in Fig. 1(d).

For the stress condition, all of the outlet branches experienced large pressure drops at peak flow (Fig. 2) indicating higher cardiac workload on all the supra-aortic vessels.

For the stress condition, the lowest pressure appears at the aortic arch, rather than the descending outlet as in the rest condition case (compared in Figs. 2 and 3). Flow separations also occur in the region around the arch and on the diverging wall downstream the site of the coarctation in the descending aorta (Fig. 4). However, due to the higher flow velocity under stress condition, the separated boundary layer grows along one side of the descending aorta wall. On the other side of the wall, flow retains higher velocity and no separation is observed.

Tables 1 and 2 summarize the average values of flow rates and pressures over one cardiac cycle. Throughout the cardiac cycle, the flow splitting ratios among the three supra-aortic vessels were assumed constant and estimated by the supplied average flow rates through each upper branch. Considering the fact that this simple boundary condition of constant flow ratio is not physiologically accurate, the effect of flow splits on the coarctation pressure gradients was further investigated.

Table 2. CFD-predicted pressure gradients (mmHg): $dp = P_{Proximal} - P_{Distal}$

		Rest			Stress		
		Min.	Max.	Mean	Min.	Max.	Mean
$P_{Proximal}$	Fine mesh	49	81	63	29	111	60
P_{Distal}	Fine mesh	48	73	60	-38	89	47
dp	Fine mesh	-7	21	2.8	-12	90	14
	measured (released at STACOM13)		6.50	1.23		45.74	14.65
dp	flow split test cases on coarse mesh						
	Inno 25%,LCC 5%, LS 11% of AscAo				-12	96	15
	evenly split				-12	115	17
	through LCC only				-12	138	13

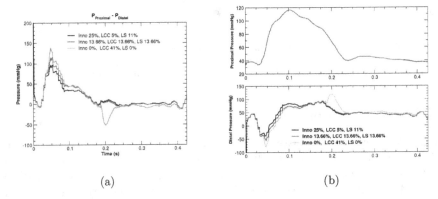

(a)   (b)

Fig. 5. (a) Pressure gradient variation with different supra-aortic flow splits. (b) Pressure (mmHg) at proximal and distal locations.

4.3   Effect of Flow Splits on Pressure Prediction

At any given instant of time, specific flow splits into the innominate, left carotid, and left subclavian arteries remain unknown. It should be noted that, other than non-uniformity of flow variables at inlets and outlets, this was the only remaining degree of freedom provided by the problem constraints. Since pressure gradients are highly dependent on the supra-aortic flow rates, the effects of different supra-aortic flow splits on predicted pressure gradients in the aorta were investigated. To this end, two additional simulations were performed for the stress condition on the coarse mesh, namely, a condition that assumed supra-aortic flow evenly split among the three upper branches; and a condition that arbitrarily assumed all supra-aortic flow is shunted only through the left carotid artery with the other two branches 100% blocked.

Fig. 5 shows the predicted pressure gradients corresponding to the supplied flow waveforms with different supra-aortic flow splits. The proximal plane is located upstream of the upper branches and close to the ascending inlet where the supplied pressure waveform is imposed. Hence, the predicted pressure at the proximal plane is little affected by the flow splitting at the upper branches. However, for distal locations at the arch and beyond, the effects of the different flow splitting ratios are stronger as evidenced by the larger pressure differences during the instances of peak flow. At lower flow rates, the splitting influences are limited. Overall, the influence of flow splitting on pressure gradients over time is moderate as indicated by the small differences in time-averaged mean values summarized in Table 2.

5   Conclusions

Unsteady flow simulations of a patient-specific aortic coarctation were conducted based on the supplied flow conditions using ANSYS Fluent software. For simplification purposes, laminar Newtonian flow, rigid aortic walls, and traditional CFD outlet BCs were assumed. Physiology-based boundary conditions and fluid-structure interactions were not considered, which challenges the accuracy of the present results. At the end of the 4th International Workshop STACOM 2013, the measured pressure gradients were released. Compared to the measured pressure gradients (shown in Table 2), peak pressure gradients are dramatically overestimated in current work. Future studies need to systematically identify and correct deficiencies of the present simplified CFD approach.

Extraction of Cardiac and Respiratory Motion Information from Cardiac X-Ray Fluoroscopy Images Using Hierarchical Manifold Learning

Maria Panayiotou[1], Andrew P. King[1], Kanwal K. Bhatia[3], R. James Housden[1], YingLiang Ma[1], C. Aldo Rinaldi[1,2], Jas Gill[1,2], Michael Cooklin[1,2], Mark O'Neill[1,2], and Kawal S. Rhode[1]

[1] Division of Imaging Sciences and Biomedical Engineering,
King's College London, UK
[2] Department of Cardiology, Guy's & St. Thomas'
Hospitals NHS Foundation Trust,
London, UK
[3] Biomedical Image Analysis Group,
Department of Computing,
Imperial College London,
London, UK
maria.panayiotou@kcl.ac.uk

Abstract. We present a novel and clinically useful method to automatically determine the regions that carry cardiac and respiratory motion information directly from standard mono-plane X-ray fluoroscopy images. We demonstrate the application of our method for the purposes of retrospective cardiac and respiratory gating of X-ray images. Validation is performed on five mono-plane imaging sequences comprising a total of 284 frames from five patients undergoing radiofrequency ablation for the treatment of atrial fibrillation. We established end-inspiration, end-expiration and systolic gating with success rates of 100%, 100% and 95.3%, respectively. This technique is useful for retrospective gating of X-ray images and, unlike many previously proposed techniques, does not require specific catheters to be visible and works without any knowledge of catheter geometry.

1 Introduction

Electrophysiology (EP) procedures are minimally invasive catheter procedures that are used to treat cardiac arrhythmias. They are carried out under X-ray fluoroscopic image guidance. However, X-ray images have poor soft tissue contrast. To overcome the lack of soft tissue contrast, pre-procedural three-dimensional (3D) images can be registered and overlaid in real-time with the 2D X-ray images using specialized hybrid imaging systems [1]. Achieving the registration in a conventional mono-plane catheter laboratory is challenging. A solution was proposed in [2] where catheters were used to constrain the registration. An implementation of this approach used 3D

O. Camara et al. (Eds.): STACOM 2013, LNCS 8330, pp. 126–134, 2014.

catheter reconstructions from sequential bi-plane X-ray images [3]. This technique required manual cardiac and respiratory phase matching of the bi-plane images. A similar requirement for automatic frame matching exists when catheter positional information needs to be measured with reference to a registered anatomical model. This is important for recording the position of electrical measurements, pacing locations and ablation treatments [4,5].

A cardiac electrocardiogram (ECG) synchronously recorded with X-ray images can be used for cardiac gating. However, this facility is not always present in X-ray systems and when present, there may be unknown delay between the ECG signal and the X-ray data. Respiratory gating can be achieved using breath-holding. This is commonly used during magnetic resonance imaging (MRI) [6]. However, this is impractical in the catheter laboratory where patients can be heavily sedated.

Image-based motion estimation should be more reliable and robust. This approach do not require any special hardware, fiducial markers or additional contrast agent. In [7,8] diaphragm tracking was used for respiratory phase determination. However, the diaphragm is not always visible in cardiac X-ray images due to collimation to reduce radiation dose. A more promising approach is to track the EP catheters [9, 10, 11]. Nevertheless, decoupling the cardiac from the respiratory motion can be challenging. The technique presented in [12] was used to estimate the motion between successive frames using a phase correlation algorithm, without requiring specific catheters to be present. The technique was tested on 2D X-ray angiographic and 3D liver and intra-cardiac ultrasound sequences. The main drawback of this technique is that it assumes that objects in the scene exhibit only translations, which is not the case in EP images.

Manifold learning (ML) has been shown to be a useful image-based method for cardiac and respiratory phase detection. In [13] the Laplacian Eigenmaps (LE) method was used for respiratory gating in MRI and ultrasound (US) applications. In US applications, LE has been used for cardiac gating [14]. In [15] a technique called Hierarchical Manifold Learning (HML) was used to learn the regional correlations in motion within a sequence of time-resolved MR images of the thoracic cavity.

We propose a novel approach for cardiac and respiratory phase gating for cardiac EP X-ray images based on HML. The algorithm is validated using X-ray images taken during radiofrequency ablation (RFA) procedures for patients being treated for atrial fibrillation (AF). The novelty of our approach is that it does not rely on specific catheters being present in the image data or the localisation of these devices, and makes no assumptions about the nature of the motion present in the images.

2   Methods

A block diagram of the proposed method for cardiac and respiratory gating is shown in Fig. 1. We first describe the respiratory gating approach and then outline how this technique is expanded with a number of other image processing operations to make it suitable for cardiac motion gating.

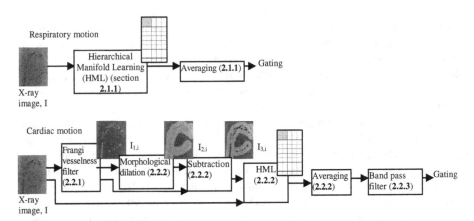

Fig. 1. Block diagram of the proposed HML-based method. (Top) For respiratory gating; (bottom) for cardiac gating.

2.1   Respiratory Gating

2.1.1   Hierarchical Manifold Learning

ML is a non-linear dimensionality reduction technique, which aims to embed data that originally lies in a high dimensional (high-D) space into a lower dimensional (low-D) space, while preserving characteristic properties. LE is a particular ML technique, which is often chosen for medical imaging applications [16]. One of the disadvantages of applying conventional ML techniques to medical images is that the whole of the image is represented by a single value even though not all the image contains relevant information. To prevent this weakness we used HML, a recently proposed technique [15]. To avoid the need to pre-define regions of interest, the images are separated into regular patches, of size 2x2 pixels each, finding a manifold embedding for each image patch. The idea is to align each patch embedding to several parent manifolds, with the strength of the aligning constraint dependent on the distance to parent patch centers. For a 2D image, it would seem natural to constrain a new patch to be close to its four nearest (in terms of distance) parent patches [15]. Consequently, for our experiments, we simply use four (2D) parent patches. As we are trying to recover both the cardiac and respiratory motions, we reduce the data to 2 dimensions. By using a manifold embedding dimensionality of 2, each patch in each time frame is represented by a 2D coordinate. We treated the values in each dimension separately. We found by visual inspection that the coordinates of the 1st dimension represent the respiratory motion while the coordinates of the 2nd dimension represent the cardiac motion. We denote the output of the HML process by P_1, P_2 for the 1st and 2nd dimension, respectively.

2.1.2   Gating

HML is applied to the 2x2 patches of our X-ray images, denoted I. Then, Eq. 1 is used to obtain the respiratory phase.

$$\bar{P}_{1,i} = \sum_{j \in I} P_{1,j,i} / N_I \tag{1}$$

where $P_{1,j,i}$ is the 1st dimension of HML result at patch j, frame i and N_I is the total number of patches in I. The peaks of the plots at each time frame represent end-inspiration (EI) respiratory frames, Ω_{IX}, while the troughs represent end-expiration (EX) respiratory frames, Ω_{EX}.

$$\Omega_{EI} = \left\{ i | \bar{P}_{1,i-1} < \bar{P}_{1,i} > \bar{P}_{1,i+1} \right\} \tag{2}$$

$$\Omega_{EX} = \left\{ i | \bar{P}_{1,i-1} > \bar{P}_{1,i} < \bar{P}_{1,i+1} \right\} \tag{3}$$

2.2 Cardiac Gating

The embedding coordinates of the 2nd dimension, P_2, of each patch for every frame of each sequence relate to the cardiac motion. However, this is significantly affected by respiratory motion. To compensate for the respiratory motion three additional steps are applied to our X-ray images, a Frangi vesselness (FV) filter [17] followed by morphological opening, morphological dilation and a band pass filter.

2.2.1 Frangi Vesselness Filter

The FV filter is applied to all X-ray images in the sequence. This technique identifies tubular structures in the X-ray images, which are expected to carry useful cardiac and respiratory motion information, using Hessian eigenvalues. The responses of the FV filter are binarised by applying a threshold. To remove the noise present while preserving the shape and size of the detected structures we apply morphological opening to the binarised responses. We denote the results of this opening process by $I_{1,i}$, where i is the X-ray frame number.

2.2.2 Morphological Dilation

The previous step is followed by the application of morphological dilation. This operation adds pixels to the boundaries of detected structures, which produces $I_{2,i}$. In morphological dilation the value of the output pixel is the maximum value of all the pixels in the input pixel's neighbourhood. Computing $I_{3,i} = I_{2,i} - I_{1,i}$ the image patches around the detected tubular structures are identified. The HML is applied to these patches only. These patches were found to be more useful for extracting cardiac motion information than using the tubular structures themselves. Although the tubular structures are expected to carry useful cardiac motion information, they also carry respiratory motion information that adversely affects the accuracy of our systolic peaks. For cardiac motion we use

$$\bar{P}_{2,i} = \sum_{j \in I_{3,i}} P_{2,j,i} / N_{I_{3,i}} \tag{4}$$

where $P_{2,j,i}$ is the 2nd dimension of the HML results at patch j, frame i and $N_{I_{3,i}}$ is the total number of patches in $I_{3,i}$.

2.2.3   Band Pass Filter and Gating

To remove residual respiratory motion, a 2nd order band pass filter is applied to our plots. The peaks of the plots represent systolic frames.

$$\Omega_{sys}=\left\{i|\bar{P}_{2,i-1}<\bar{P}_{2,i}>\bar{P}_{2,i+1}\right\}\tag{5}$$

3   Experiments

3.1   Materials

All patient procedures were carried out using a mono-plane 25cm-flat-panel cardiac X-ray system (Philips Allura Xper FD10, Philips Healthcare, Best, The Netherlands), in one of the catheterization laboratories at St. Thomas' Hospital, London, U.K. In total, 5 different clinical fluoroscopy sequences from 5 patients who underwent RFA procedures for the treatment of AF were used. A total of 284 X-ray images were processed. For each patient, X-ray imaging was performed at 3 frames per second. All X-ray images were 512×512 pixels in resolution, with a pixel to mm ratio of 0.25. Included in this ratio is the typical magnification factor of the X-ray system.

3.2   Optimization of Parameters

We built our algorithm using the leave-one-out cross-validation (LOOCV) approach. This involved using 4 sequences as the training data and the remaining sequence as the validation data. To build our algorithm, we optimised the four parameters involved in extracting the cardiac phases. These parameters include: a threshold level on the normalised output, 0 to 1, of the FV filter, the number of morphological dilations of the identified structures, and the pass band and stop band frequencies of the band pass filter, optimised to be >0.02, 30±3, 0.62 and 0.98, respectively.

3.3   Validation of Our Retrospective Cardiac and Respiratory Gating

We validated the respiratory gating using either diaphragm or heart border tracking as described in [8] for the ground truth. The choice of ground truth was determined by which structure was visible in the X-ray images. The signals obtained using the tracking method (gold standard) were compared to the signals obtained using the HML-based method. In order to validate our cardiac gating method, manual gating of the cardiac cycle at systole was performed by an experienced observer, by visually detecting the onset of contraction of the left ventricle from the fluoroscopic left heart border shadow. The systolic frame number was recorded and compared with the corresponding systolic frame number from the automatic detection. We chose systole as opposed to diastole for validation since the manual ground truth is more reliable for systole where rapid motion can be used as the visual cue.

4   Results

4.1   FV Filter and Morphological Dilation Output Images

Fig. 2(a) gives an illustration of the output of the thresholded FV filter response, $I_{1,1}$, of the first frame of one example X-ray sequence after the application of the threshold level and the morphological opening, overlaid with the corresponding X-ray image. Fig. 2(b) illustrates the image output, $I_{3,1}$, overlaid with the corresponding X-ray image for the first frame of the same example case.

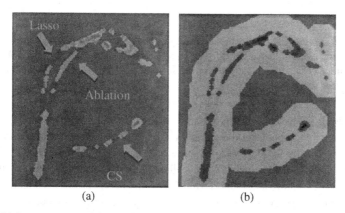

(a)   (b)

Fig. 2. (a) Image output of the FV filter followed by morphological opening, $I_{1,1}$, overlaid with the corresponding X-ray image for Case 1. The EP catheters during the ablation stage of a procedure are shown. (b) Image output, $I_{3,1}$, overlaid with the corresponding X-ray image for Case 1.

4.2   Qualitative Validation

For respiratory gating validation, a plot of the respiratory trace obtained using the HML-based method for Case 1 is illustrated in Fig. 3(a) as a solid black line. The diaphragm/heart border tracking is shown as a dashed black line. The results of the cardiac gating validation are shown in Fig. 3(b) for the first 45 frames for the same case. The plotted vertical black lines correspond to the gold standard systolic frames.

4.3   Comparative Quantitative Validation

It is important to investigate whether our new HML-based technique is superior to our previously presented retrospective PCA-based gating approach [11]. Therefore, for both cardiac and respiratory gating, the absolute frame difference was computed between the HML-based method and the gold standard methods. Specifically, systolic, EI and EX frames were recorded from the HML-based and gold standard methods and their corresponding absolute frame differences were computed. This was

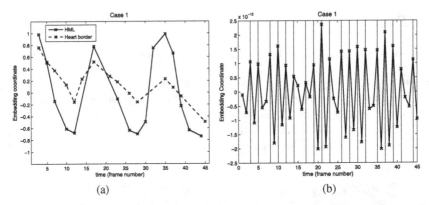

(a)   (b)

Fig. 3. (a) Graphical representation of the obtained respiratory trace with X-ray frame number. The heart border tracking (gold standard) is shown as a dashed black line. (b) The HML-based method cardiac trace obtained is illustrated for the same example case. The vertical black lines are the gold standard identification of systole.

also done for the PCA-based technique. Faultless gating results are signified when the absolute frame difference between the automatic and gold standard method is zero. The results can be seen in the frequency distribution bar charts in Fig. 4(a, b and c), for EI, EX and systolic gating, respectively. Results illustrate that our HML-based method is faultless in EI and EX gating and outperforms the results of our PCA-based technique. It is also almost faultless for systolic gating. While our PCA-based technique outperforms the HML-based technique in cardiac gating, it relies on the tracking of a specific catheter, the CS catheter. Our proposed HML-based technique does not depend on any particular catheter being present in our X-ray images and requires no knowledge of catheter geometry.

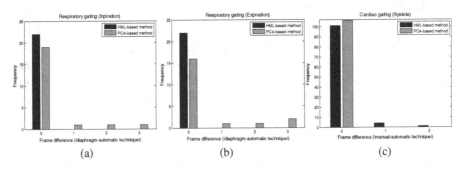

(a)   (b)   (c)

Fig. 4. Frequency distributions of frame difference errors for (a) End-inspiration gating (b) End-expiration gating and (c) Cardiac gating. Results are illustrated for the HML-based method (blue) and PCA-based method (green)

Percentage success rates were computed using 100 (x/x_{total}), where x corresponds to the number of perfectly matched gold standard and automatic gating frames and x_{total} corresponds to the total number of gold standard gating frames. Percentage success rates computed for EI, EX and systolic gating for our proposed HML-based technique were calculated to be 100%, 100% and 95.3%, respectively. For cardiac gating 4 extra false systolic peaks were obtained over all processed sequences.

5 Discussion and Conclusions

We have presented a novel and robust retrospective HML-based method for image-based automatic cardiac and respiratory motion gating. This method is able to detect cardiac and respiratory phase directly from X-ray images. We have applied our technique on 5 clinical fluoroscopy sequences and computed the success rates for EI, EX and systolic gating which were 100%, 100% and 95.3%, respectively. The HML-based method is fully automatic, requires no user interaction, no prior knowledge and can operate within a few seconds per image sequence. As our technique is not dependent on any particular catheter being present in the procedure, it has potential application in more types of cardiac catheterization procedures, rather than only RFA procedures. The method will be particularly useful for registration and overlay of pre-procedural images with X-ray fluoroscopy for guidance and biophysical modelling. Future work will focus on testing our method on X-ray coronary angiography images where no catheters are present. Investigations show that the reason for optimising systolic gating on the specified patches is because of the inclusion of the heart border, a structure that carries significant cardiac motion information. Further work will focus on modifying our technique by giving more emphasis to the importance of the heart border structure in an attempt to improve the systolic gating success rate.

Acknowledgement. This work is funded by EPSRC programme grant EP/H046410/1.

References

1. Rhode, K.S., Hill, D.L.G., Edwards, P.J., Hipwell, J., Rueckert, D., Sanchez-Ortiz, G., Hegde, S., Rahunathan, V., Razavi, R.: Registration and tracking to integrate X-ray and MR images in an XMR facility. IEEE Trans. Med. Imaging 22(11), 1369–1378 (2003)
2. Sra, J., Narayan, G., Krum, D., Malloy, A., Cooley, R., Bhatia, A., Dhala, A., Blanck, Z., Nangia, V., Akhtar, M.: Computed tomography-fluoroscopy image integration-guided catheter ablation of atrial fibrillation. Cardiovasc. Electrophysioly 18, 409–414 (2007)
3. Truong, M.V., Gordon, T., Razavi, R., Penney, G.P., Rhode, K.S.: Analysis of catheter-based registration with vessel-radius weighting of 3D CT data to 2D X-ray for cardiac catheterisation procedures in a phantom study. Statistical Atlases and Computational Models of the Heart Imaging and Modelling Challenges, 139–148 (2012)
4. Rhode, K.S., et al.: A System for Real-Time XMR Guided Cardiovascular Intervention. IEEE Trans. Med. Imaging 24(11), 1428–1440 (2005)

5. Sermesant, M., et al.: Patient-specific electromechanical models of the heart for the prediction of pacing acute effects in CRT: a preliminary clinical validation. Med. Image Anal. 16(1), 201–215 (2012)
6. Paling, M.R., Brookeman, J.R.: Respiration artifacts in MR imaging: reduction by breath holding. J. Comput. Assist. Tomogr. 10(6), 1080–1082 (1986)
7. Shechter, G., Shechter, B., Resar, J.R., Beyar, R.I.: Prospective motion correction of X-ray images for coronary interventions. IEEE Trans. Med. Imaging 24, 441–450 (2005)
8. Ma, Y.L., King, A.P., Gogin, N., Gijsbers, G., Rinaldi, C.A., Gill, J., Razavi, R.S., Rhode, K.S.: Clinical evaluation of respiratory motion compensation for anatomical roadmap guided cardiac electrophysiology procedures. IEEE Trans. Biomed. Eng. 59(1), 122–131 (2012)
9. Brost, A., Wimmer, A., Bourier, F., Koch, M., Liao, R., Kurzidim, K., Strobel, N., Hornegger, J.: Constrained registration for motion compensation in atrial fibrillation ablation procedures. IEEE Trans. Med. Imaging 31(4), 870–881 (2012)
10. Ma, Y.L., King, A.P., Gogin, N., Rinaldi, C.A., Gill, J., Razavi, R., Rhode, K.S.: Real-time respiratory motion correction for cardiac electrophysiology procedures using image-based coronary sinus catheter tracking. Med. Image Comput. Assist. Interv. 13, 391–399 (2010)
11. Panayiotou, M., King, A.P., Ma, Y.L., Rinaldi, C.A., Gill, J., Cooklin, M., O'Neill, M., Rhode, K.S.: Automatic image-based retrospective gating of interventional cardiac X-ray images. In: Conf. Proc. IEEE Eng. Med. Biol. Soc., pp. 4970–4973 (2012)
12. Sundar, H., Khamene, A., Yatziv, L., Xu, C.: Automatic image-based cardiac and respiratory cycle synchronization and gating of image sequences. In: Yang, G.-Z., Hawkes, D., Rueckert, D., Noble, A., Taylor, C. (eds.) MICCAI 2009, Part II. LNCS, vol. 5762, pp. 381–388. Springer, Heidelberg (2009)
13. Yigitsoy, C., Rijkhorst, E., Navab, N., Wachinger, C.: Manifold learning for image based breathing gating in ultrasound and MRI. Medical Image Analysis 16(4), 806–818 (2012)
14. Isguder, G.G., Unal, G., Groher, M., Navab, N., Kalkan, A.K., Degertekin, M., Hetterich, H., Rieber, J.: Manifold learning for image-based gating of intravascular ultrasound (IVUS) pullback sequences. In: Liao, H., "Eddie" Edwards, P.J., Pan, X., Fan, Y., Yang, G.-Z., et al. (eds.) MIAR 2010. LNCS, vol. 6326, pp. 139–148. Springer, Heidelberg (2010)
15. Bhatia, K.K., Rao, A., Price, A.N., Wolz, R., Hajnal, J., Rueckert, D.: Hierarchical manifold learning. In: Ayache, N., Delingette, H., Golland, P., Mori, K. (eds.) MICCAI 2012, Part I. LNCS, vol. 7510, pp. 512–519. Springer, Heidelberg (2012)
16. Belkin, M., Niyogi, P.: Laplacianeigenmaps for dimensionality reduction and data representation. Neural Computation 15(6), 1373–1396 (2003)
17. Frangi, A.F., Niessen, W.J., Vincken, K.L., Viergever, M.A.: Multiscale vessel enhancement filtering. In: Wells, W.M., Colchester, A.C.F., Delp, S.L. (eds.) MICCAI 1998. LNCS, vol. 1496, pp. 130–137. Springer, Heidelberg (1998)

Dyadic Tensor-Based Interpolation of Tensor Orientation: Application to Cardiac DT-MRI*

Jin Kyu Gahm[1,2] and Daniel B. Ennis[2]

[1] Department of Computer Science, UCLA, CA 90095, USA
[2] Department of Radiological Sciences, UCLA, CA 90095, USA
gahmj@ucla.edu

Abstract. *Objective*: To develop an accurate and mathematically un-ambiguous method for interpolation of tensor orientation, specifically for the interpolation of cardiac microstructural orientation. *Methods*: A dyadic tensor-based (DY) orientation interpolation method, which sidesteps the eigenvector sign ambiguity problem by interpolating between the dyadic tensors of eigenvectors, is proposed and evaluated. The quaternion-based (QT) orientation interpolation method, which interpolates along the minimum rotation path between tensor orientations, is also revised and evaluated. DY and QT are compared to conventional tensor-based interpolation methods using both synthetic and cardiac DT-MRI data. *Results*: All methods (except QT) perform similarly well for recovery of the primary eigenvector. DY has significantly less bias than all other methods for recovery of the secondary and tertiary eigenvector, which is especially important for interpolating myolaminar sheet orientation. *Conclusion*: DY is a fast, commutative, and mathematically unambiguous tensor orientation interpolation method that accurately interpolates cardiac microstructural orientation.

1 Introduction

Diffusion tensor magnetic resonance imaging (DT-MRI) [1] characterizes soft tissue microstructural organization by measuring, for example, tissue anisotropy and myofiber and myolaminae orientations. DT-MRI methods estimate the self-diffusion tensor of water in each image voxel. The second-order symmetric positive definite diffusion tensor ($\mathbf{D}$) can be decomposed into eigenvalues (λ_i, shape) and eigenvectors ($\mathbf{e}_i$, orientation). Tensor shape can also be intuitively and saliently represented by tensor invariants such as tensor trace (J_1), fractional anisotropy (FA, J_2) and tensor mode (J_3) [2–5].

The three eigenvectors correspond to the myofiber long-axis ($\mathbf{e}_1$), the cross-fiber direction within the myolaminar sheet ($\mathbf{e}_2$) and the sheet-normal direction ($\mathbf{e}_3$) in cardiac applications [6]. To build computational models of cardiac mechanics and electrophysiology (EP), both myofiber and myolaminae orientation

---

* This work was supported, in part, by grant support from the NIH (P01 HL78931) and the Department of Radiological Sciences at UCLA.

information is required at millions of closely spaced nodes. DT-MRI measurements, however, are on a lattice and typically number $< 1e6$ for *ex vivo* studies $< 1e4$ for *in vivo* studies [7], so interpolation of tensor orientation is needed. The orientation (SO(3)) interpolation problem has been widely studied in the computer graphics literature, but the tensor orientation interpolation problem in DT-MRI is more challenging because eigenvectors have an arbitrary sign (physiologically and mathematically) so tensor orientation cannot be uniquely described.

Most of the conventional approaches have been tensor-based and amongst the simplest is the Euclidean (EU) method, but it suffers from the tensor shape swelling effect [8, 9]. The affine-invariant Riemannian (AI) and log-Euclidean (LE) tensor interpolation methods [8, 9] were proposed to solve the tensor shape (tensor swelling) problem, but they underestimate other tensor invariants including tensor trace and FA [5, 10]. The geodesic-loxodrome (GL) method [4] guarantees monotonic interpolation of orthogonal tensor invariants [2], but is computationally expensive. The linear invariant (LI) method [5] linearly interpolates tensor invariants (shape) at significantly reduced computational cost, but no new method for tensor orientation interpolation was presented. The tensor-based methods mostly focus on tensor shape interpolation, and no distinct advantage of the methods in tensor orientation has been reported [5]. Recently a separate tensor interpolation method [10] was proposed that interpolates Euler angles or quaternions along the minimum rotation path between tensor orientations, but it was not quantitatively validated using cardiac DT-MRI data.

We propose a new dyadic-tensor based (DY) tensor orientation interpolation method that sidesteps the eigenvector sign ambiguity problem by interpolating between the dyadic tensors of eigenvectors with subsequent reduction to rank-1 dyadics and orthogonal matrices. We also revise and simplify the quaternion-based (QT) method [10], and evaluate it using cardiac DT-MRI data. The QT and DY tensor-based methods are compared to the tensor-based interpolation methods including EU, AI, LE and GL for accurate recovery of cardiac microstructural orientation using four experimentally measured DT-MRI datasets from rabbit and pig hearts.

2   Theory

Quaternion-Based Interpolation. One approach to resolve the eigenvector sign ambiguity problem is to directly tackle it by choosing the minimum rotation path between tensor orientations. Tensor orientation is commonly represented by a rotation matrix $\mathbf{R} = [\mathbf{e}_i]$ consisting of three eigenvectors, sorted in descending order of their corresponding eigenvalues, but can also be represented by a unit quaternion $\mathbf{q} = a + bi + cj + dk = [a, b, c, d]$ where $a^2 + b^2 + c^2 + d^2 = 1$. Tensor orientation has four different descriptions intuitively represented by rotation matrices $\mathbf{RP}$ where $\mathbf{P} = \text{diag}(p_j)$ such that $p_j = \pm 1$ and $p_1 p_2 p_3 = 1$, which can be converted into unit quaternions $\mathbf{q}_k$:

$$\mathbf{q}_k = [a, b, c, d], [b, -a, d, -c], [c, -d, -a, b], [d, c, -b, -a] . \qquad (1)$$

Then the minimum rotation path between two tensor orientations $\mathbf{R_A}$ and $\mathbf{R_B}$ can be determined by the maximum magnitude of inner products between fixed $\mathbf{q_A}$ and four different $\mathbf{q_B}$ (or between fixed $\mathbf{q_B}$ and four different $\mathbf{q_A}$). If the maximum value has a negative sign, the corresponding quaternion $\mathbf{q_B}$ (or $\mathbf{q_A}$) should be negated. Once the unit quaternions are uniquely determined, normalized linear interpolation (nlerp) is used:

$$\mathbf{q_C} = \left((1 - t)\mathbf{q_A} + t\mathbf{q_B}\right) / \|(1 - t)\mathbf{q_A} + t\mathbf{q_B}\| , \tag{2}$$

which is computationally less expensive than spherical linear interpolation (slerp) [11]. The interpolated quaternion $\mathbf{q_C}$ is easily converted to a rotation matrix $\mathbf{R_C}$.

Dyadic Tensor-Based Interpolation. Our approach is to sidestep the sign ambiguity problem by using dyadic tensors [12]. Dyadic tensors of eigenvectors $\mathbf{e}_i$ are defined by $\mathbf{E}_i = \mathbf{e}_i \otimes \mathbf{e}_i = \mathbf{e}_i\mathbf{e}_i^T$. Note, $\mathbf{e}_i \otimes \mathbf{e}_i = -\mathbf{e}_i \otimes -\mathbf{e}_i$. Dyadic tensors are rank-1 with only one non-zero eigenvalue whose value is 1 and the corresponding eigenvector is exactly $\mathbf{e}_i$ or $-\mathbf{e}_i$. Interpolation between $\mathbf{R_A} = [\mathbf{e}_{\mathbf{A}i}]$ and $\mathbf{R_B} = [\mathbf{e}_{\mathbf{B}i}]$ starts with linear interpolation between their dyadic tensors:

$$\mathbf{F}_i = (1 - t)\mathbf{E}_{\mathbf{A}i} + t\mathbf{E}_{\mathbf{B}i} . \tag{3}$$

Since $\mathbf{F}_i$ are not generally rank-1, the nearest rank-1 dyadic tensor $(\mathbf{x} \otimes \mathbf{x})$ can be obtained by minimizing:

$$J(\mathbf{x}) = \|\mathbf{F}_i - \mathbf{x} \otimes \mathbf{x}\|_F^2 = \mathrm{tr}\left\{(\mathbf{F}_i - \mathbf{x}\mathbf{x}^T)^T(\mathbf{F}_i - \mathbf{x}\mathbf{x}^T)\right\}$$
$$= \mathrm{tr}\left\{\mathbf{F}_i^2 - 2\mathbf{F}_i\mathbf{x}\mathbf{x}^T + (\mathbf{x}\mathbf{x}^T)^2\right\} = \|\mathbf{F}_i\|_F^2 - 2\mathrm{tr}(\mathbf{x}^T\mathbf{F}_i\mathbf{x}) + \|\mathbf{x}\|^4 , \tag{4}$$

where $\|\cdot\|_F$ denotes the Frobenius norm, and the derivative is:

$$J'(\mathbf{x}) = -4\mathbf{F}_i\mathbf{x} + 4\|\mathbf{x}\|^2\mathbf{x} . \tag{5}$$

By setting the derivative equal to zero, the eigenvalue equation $\mathbf{F}_i\mathbf{x} = \|\mathbf{x}\|^2\mathbf{x}$ is obtained. Therefore, the eigenvector $\mathbf{m}_i$ corresponding to the largest eigenvalue of $\mathbf{F}_i$ minimizes Eq. 4. However, since the interpolation between dyadic tensors is separately performed on each pair of eigenvectors the matrix $\mathbf{M} = [\mathbf{m}_i]$ is not generally orthogonal. The orthogonal matrix closest to $\mathbf{M}$ can be obtained by minimizing:

$$w_1\|\mathbf{x}_1 - \mathbf{m}_1\|^2 + w_2\|\mathbf{x}_2 - \mathbf{m}_2\|^2 + w_3\|\mathbf{x}_3 - \mathbf{m}_3\|^2 , \tag{6}$$

where $[\mathbf{x}_i]$ is an orthogonal matrix, and w_i are the eigenvalues computed by the LI method [5], which assigns different weights to each eigenvector term according to the interpolated tensor shape. Equation 6 can be rewritten in a matrix form:

$$\|(\mathbf{M} - \mathbf{X})\mathbf{W}\|_F^2 = \mathrm{tr}\left\{(\mathbf{MW} - \mathbf{XW})(\mathbf{MW} - \mathbf{XW})^T\right\}$$
$$= \mathrm{tr}\left\{(\mathbf{MW})(\mathbf{MW})^T\right\} + \mathrm{tr}\left(\mathbf{XWW}^T\mathbf{X}^T\right) - 2\mathrm{tr}\left(\mathbf{MW}^2\mathbf{X}^T\right)$$
$$= \|\mathbf{MW}\|_F^2 + \|\mathbf{W}\|_F^2 - 2\mathrm{tr}\left(\mathbf{MW}^2\mathbf{X}^T\right) , \tag{7}$$

where $\mathbf{X} \in O(3)$ and $\mathbf{W}^2 = \text{diag}(w_i)$. Minimizing Eq. 7 is achieved by maximizing:

$$\text{tr}\left(\mathbf{MW}^2\mathbf{X}^T\right) = \text{tr}\left(\mathbf{U\Sigma V}^T\mathbf{X}^T\right) = \text{tr}\left(\mathbf{V}^T\mathbf{X}^T\mathbf{U\Sigma}\right) \leq \text{tr}\left(\mathbf{\Sigma}\right) , \qquad (8)$$

where $\mathbf{U}$, $\mathbf{\Sigma}$ and $\mathbf{V}$ are obtained from the singular value decomposition (SVD) of $\mathbf{MW}^2 = \mathbf{U\Sigma V}^T$, implying that Eq. 8 is maximized when $\mathbf{V}^T\mathbf{X}^T\mathbf{U} = \mathbf{I} \Leftrightarrow \mathbf{X} = \mathbf{UV}^T$. Therefore, the interpolated tensor orientation $\mathbf{R_C} = [\mathbf{e_{C}}_i]$ can be obtained by replacing the singular values with ones from the SVD of $\mathbf{MW}^2$. If the determinant of $\mathbf{R_C}$ is -1, then $\mathbf{R_C}$ should be negated to be a right-handed rotation matrix.

3   Methods

Synthetic Tensors. Using the EU, LE, GL, quaternion-based (QT) and dyadic tensor-based (DY) methods, interpolation was performed between two tensors of the same shape ($J_i = \{1, 0.5, 0.8\}$), and different orientations whose angles between each pair of eigenvectors are $82°$, $45°$ and $64°$. LI was used for tensor shape interpolation and combined with QT and DY for complete tensor interpolation.

Real DT-MRI Data. The rabbit heart DT-MRI data was acquired using a 7T Bruker Biospin scanner, and a 3D fast spin echo sequence with the following imaging parameters: TE/TR = 30/500 ms, b-value = 1000 s/mm^2, 24 diffusion gradient encoding directions, 6 nulls, and RARE factor two. The in-plane imaging resolution was 0.5×0.5×0.80 mm obtained by using a 96×96 encoding matrix, 72–96 slices and a 48×48×54–72 mm imaging volume. The pig heart DT-MRI data was acquired using a Siemens 1.5T Avanto and a 3T Trio scanner, and a 2D readout-segmented echo-planar pulse sequence with the following imaging parameters: TE/TR = 80/6800 ms, b-value = 1000 s/mm^2, 30 diffusion gradient encoding directions, one null, 15 readout segments, and 8-10 averages. The in-plane imaging resolution was 1×1×3 mm obtained by using an 150×150 encoding matrix, 43–44 slices and a 150×150×129–132 mm imaging volume. Diffusion tensors were estimated without zero padding and with linear regression.

Evaluation Procedure. The same tensor orientation evaluation procedure proposed in [5] was applied to the two rabbit and two pig heart DT-MRI datasets. The median autocorrelation (AC) length for every dimension was computed in each tensor invariant (J_i) map of the segmented myocardium. The myocardial DT-MRI volume was down-sampled in each dimension by a factor of the smallest integer not less than the median AC length for each tensor invariant map, and trilinear tensor orientation interpolation was performed with the EU, AI, LE, GL, QT and DY methods at the removed voxels using the remaining data. Then the interpolated tensor orientations by each method were compared to the originally measured data by computing the angle difference between each pair of eigenvectors. Subsequently the population of the angle difference data

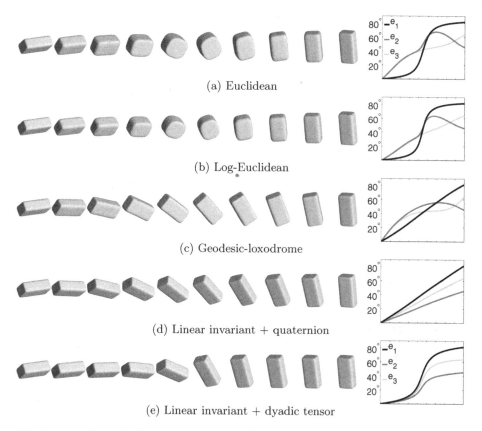

(a) Euclidean

(b) Log-Euclidean

(c) Geodesic-loxodrome

(d) Linear invariant + quaternion

(e) Linear invariant + dyadic tensor

Fig. 1. Interpolation between two synthetic tensors of equal shape and different orientation. The angle between every pair of the primary, secondary and tertiary eigenvectors is monotonically interpolated only in (d) and (e). All the tensor-based methods (a), (b), and (c) fail to monotonically interpolate the angle between the secondary eigenvectors.

was spatially decorrelated by decimating the data in every dimension by the smallest integer not less than the AC lengths, and the decorrelated data was bootstrapped 1000 times by random sampling with replacement to compute the 95% confidence interval (CI) about the median.

4   Results

Synthetic Example. Figure 1 shows an example of interpolation between two synthetic tensors with the same shape and different orientations using the EU, LE, GL, LI+QT and LI+DY methods. Tensors are visualized as superquadric glyphs [13], and plots of each eigenvector's angle relative to the leftmost tensor's orientation are shown along the interpolation paths. EU and LE fail to preserve the tensor shape during rotation, but GL, LI+QT and LI+DY maintain the

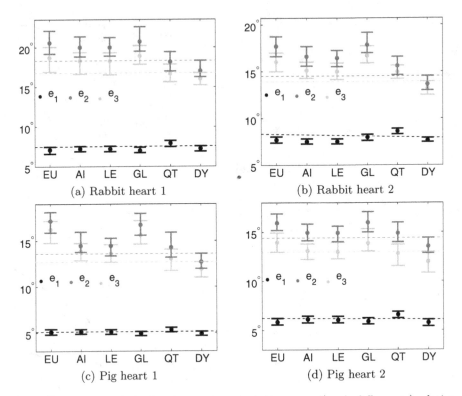

(a) Rabbit heart 1   (b) Rabbit heart 2

(c) Pig heart 1   (d) Pig heart 2

Fig. 2. Bootstrap statistics for eigenvector orientation errors (angle differences) relative to real DT-MRI data. Each dot represents the median angle difference, and each error bar represents the bootstrapped 95% CI of the median. The (black, dark gray, and light gray) dashed lines represent the upper limits of DY's CIs associated with the (primary, secondary, and tertiary) eigenvectors, which define whether or not DY's CIs overlap with the others'. DY introduces the least error to the secondary and tertiary eigenvector orientations, and similar errors to the primary eigenvector orientation compared to the tensor-based methods (EU, AI, LE and GL).

tensor shape. With respect to tensor orientation, QT and DY monotonically interpolates the angle of every eigenvector. The tensor-based methods (EU, LE and GL), however, fail to monotonically interpolate the angle of the secondary eigenvector.

DY's monotonic interpolation of each eigenvector needs to be more carefully investigated. Each method has a distinct interpolation path between tensor orientations, and QT's path is explicitly the minimum rotation path. Monotonic interpolation of eigenvectors and/or the minimum rotation path does not imply interpolation of tensor orientation with the least error. Therefore, we experimentally evaluated each method using real DT-MRI data.

Evaluation Statistics. The smallest integers not less than the median AC lengths were 2, 2, and 3 for the rabbit heart data and 3, 3, and 2 for the pig heart data in the x-, y- and z-directions, respectively. Figure 2 shows the bootstrap statistics of angle differences between each eigenvector pair from the original and interpolated tensor orientations.

Comparison of the orientation errors between methods reveals that each method performs consistently across the various data sets (e.g. errors for recovering e_1 significantly decrease from QT to DY). QT's e_1 median error, however, is significantly higher than all other methods (i.e. 95% CI does not overlap) for the rabbit data, but not for the pig data. DY performs similarly to conventional tensor interpolation methods for recovering e_1 in both rabbit and pig DT-MRI data.

DY has the lowest median error for recovery of both e_2 and e_3 compared to all other methods. Notably, DY has a significantly lower median recovery error for e_2 and e_3 compared to either EU or GL for all four datasets.

5 Conclusion

Accurate interpolation of myofiber and myolaminar sheet orientations is essential for computational modeling of cardiac mechanics and electrophysiology (EP). Cardiac mechanics and EP modeling requires accurate tensor orientation information at every computational node in order to assign correctly the axes of anisotropic electrical activation.

The comparison results show that DY performs significantly better than the tensor based methods, especially EU and GL, for recovery of each component of cardiac microstructural orientation. In particular, the improvement in recovery of the secondary and tertiary eigenvectors is important for recovery of myolaminar sheet orientation. Note that QT's minimum rotation path has significantly larger median errors for recovery of the primary eigenvector than DY's interpolation path.

LI+DY is a commutative, computationally efficient (compared to GL's numerical solution), and mathematically unambiguous tensor interpolation method that most accurately interpolates both cardiac microstructural shape [5] and orientation. Further investigations using brain DT-MRI data and the same evaluation process may be needed to evaluate if the most accurate interpolation is dependent on the underlying tissue characteristics. Furthermore, the required tensor interpolation accuracy for cardiac mechanics and EP simulations remains incompletely understood.

References

1. Basser, P.J., Mattiello, J., LeBihan, D.: Estimation of the effective self-diffusion tensor from the NMR spin echo. J. Magn. Reson. B 103(3), 247–254 (1994)
2. Ennis, D.B., Kindlmann, G.: Orthogonal tensor invariants and the analysis of diffusion tensor magnetic resonance images. Magn. Reson. Med. 55, 136–146 (2006)
3. Kindlmann, G., Ennis, D.B., Whitaker, R.T., Westin, C.F.: Diffusion tensor analysis with invariant gradients and rotation tangents. IEEE Trans. Med. Imaging 26(11), 1483–1499 (2007)

4. Kindlmann, G., San José Estépar, R., Niethammer, M., Haker, S., Westin, C.-F.: Geodesic-loxodromes for diffusion tensor interpolation and difference measurement. In: Ayache, N., Ourselin, S., Maeder, A. (eds.) MICCAI 2007, Part I. LNCS, vol. 4791, pp. 1–9. Springer, Heidelberg (2007)

5. Gahm, J.K., Wisniewski, N., Kindlmann, G., Kung, G.L., Klug, W.S., Garfinkel, A., Ennis, D.B.: Linear invariant tensor interpolation applied to cardiac diffusion tensor MRI. In: Ayache, N., Delingette, H., Golland, P., Mori, K. (eds.) MICCAI 2012, Part II. LNCS, vol. 7511, pp. 494–501. Springer, Heidelberg (2012)

6. Kung, G.L., Nguyen, T.C., Itoh, A., Skare, S., Ingels Jr., N.B., Miller, D.C., Ennis, D.B.: The presence of two local myocardial sheet populations confirmed by diffusion tensor MRI and histological validation. J. Mag. Res. Imaging 34(5), 1080–1091 (2011)

7. Tseng, W.Y., Reese, T.G., Weisskoff, R.M., Wedeen, V.J.: Cardiac diffusion tensor MRI in vivo without strain correction. Magn. Reson. Med. 42(2), 393–403 (1999)

8. Pennec, X., Fillard, P., Ayache, N.: A Riemannian framework for tensor computing. Int. J. Comp. Vis. 66, 41–66 (2006)

9. Arsigny, V., Fillard, P., Pennec, X., Ayache, N.: Log-Euclidean metrics for fast and simple calculus on diffusion tensors. Magn. Reson. Med. 56, 411–421 (2006)

10. Yang, F., Zhu, Y.M., Magnin, I.E., Luo, J.H., Croisille, P., Kingsley, P.B.: Feature-based interpolation of diffusion tensor fields and application to human cardiac DT-MRI. Med. Image Anal. 16(2), 459–481 (2012)

11. Blow, J.: Understanding slerp, then not using it. The Inner Product (April 2004)

12. Basser, P.J., Pajevic, S.: Statistical artifacts in diffusion tensor MRI (DT-MRI) caused by background noise. Magn. Reson. Med. 44, 41–50 (2000)

13. Ennis, D.B., Kindlmann, G., Rodriguez, I., Helm, P.A., McVeigh, E.R.: Visualization of tensor fields using superquadric glyphs. Magn. Reson. Med. 53, 169–176 (2005)

Continuous Spatio-temporal Atlases of the Asymptomatic and Infarcted Hearts

Pau Medrano-Gracia[1], Brett R. Cowan[1], David A. Bluemke[2],
J. Paul Finn[3], Alan H. Kadish[4], Daniel C. Lee[4], João A.C. Lima[5],
Avan Suinesiaputra[1], and Alistair A. Young[1]

[1] Dept. Anatomy with Radiology, University of Auckland, New Zealand
{p.medrano,b.cowan,a.suinesiaputra,a.young}@auckland.ac.nz
[2] NIH Clinical Ctr., Bethesda, MD, USA
bluemked@cc.nih.gov
[3] Diagnostic CardioVascular Imaging Section, UCLA, Los Angeles, CA, USA
PFinn@mednet.ucla.edu
[4] Feinberg Cardiovascular Research Inst., Northwestern University, Chicago, IL, USA
{a-kadish,dlee}@northwestern.edu
[5] The Johns Hopkins Hospital, Baltimore, MD, USA
jlima@jhmi.edu

Abstract. Statistical descriptions of regional wall motion abnormalities of the heart are key to understanding both sub-clinical and clinical progression of dysfunction. In this paper we establish a temporal registration framework of the cardiac cycle to build a spatio-temporal atlas of 300 asymptomatic volunteers and 300 symptomatic patients with myocardial infarction. A finite-element model was customised to each person's magnetic resonance images with expert-guided semi-automatic spatial and temporal registration of model parameters. A piece-wise linear temporal registration from user-defined key frames was followed by a Fourier series temporal estimation, providing temporal continuity. All spatial and temporal data were then statistically analysed by means of principal component analysis. Results show differences in sphericity, wall thickening and mitral valve dynamics between the two groups. The modes are available from www.cardiacatlas.org. These atlases can be readily applied to abnormality detection and quantification and can also aid in anatomically constrained shape-based algorithms in automatic planning or segmentation.

1 Introduction

Cardiac magnetic resonance imaging (CMRI) provides detailed spatial and functional information of the human heart. Typically, clinical parameters of interest include endocardial volume, left ventricular (LV) mass, wall thickening and ejection fraction. However, regional wall motion abnormalities are clinically important in the diagnosis and evaluation of regional heart disease such as myocardial infarction. These take the form of spatial variation of temporal characteristics

O. Camara et al. (Eds.): STACOM 2013, LNCS 8330, pp. 143–151, 2014.
© Springer-Verlag Berlin Heidelberg 2014

which are at present qualitatively assessed by clinicians as being normal or abnormal, for example in determining regional wall motion scores.

Statistical atlases of the heart are, in this context, collections of patient datasets, which can be the images themselves or derived measurements or models which have been registered to a common reference. In this paper we build the latter, i.e. a distribution of regional wall motion in terms of shape and function from two different populations. These atlases are becoming increasingly popular in both the bioengineering [5, 14] and clinical fields [11] since they offer an unprecedented quantitative comparison between a patient and a population. However, to date most atlases have typically focused on specific time points such as end-diastole (ED) and end-systole (ES) and do not usually include all temporal information [10, 12, 13]. In our previous work [14], a similar finite-element model was used however the data and methodology were different. Examples of fully spatio-temporal atlases include [3, 5, 8].

Full coverage of the time domain presents two main challenges, time registration and continuous interpolation. In this paper we address these two challenges and present an asymptomatic and symptomatic spatio-temporal atlas through their modes of variation. The main contributions of this paper are:

1. A compact representation of the spatio-temporal variation of regional wall motion in terms of a parametric model with a relatively small number of parameters
2. Application to a reasonably large number of asymptomatic and symptomatic cases
3. Identification of clinically important shape indicators including sphericity and wall thickening in the symptomatic vs. asymptomatic groups.

2   Data and Methods

Image data were obtained using the Cardiac Atlas Project [6] from two clinical studies: the Multi-Ethnic Study of Atherosclerosis (MESA) study [1] for the asymptomatic cohort and the Defibrillators To Reduce Risk By Magnetic Resonance Imaging Evaluation (DETERMINE) clinical trial [9] for the symptomatic sample. Three hundred cases were randomly selected from each study. A typical dataset comprised 20-30 frames in 6-8 short-axis slices and 3-4 long-axis slices (imaging parameters can be found in [1, 9]). All images were acquired using prospective electrocardiogram gating and therefore cover the entire cycle.

At the time of recruitment, the MESA study protocol ensured that participants did not have clinical evidence of heart attack, angina, stroke, heart failure or atrial fibrillation [1]. The DETERMINE study was designed as a prospective, multi-centre, randomised, clinical trial in patients with coronary artery disease and mild-to-moderate LV dysfunction [9].

Guide-point modelling [19] was used to adaptively optimise a time-varying 3D finite-element model of the LV to fit each subject's images using custom software (CIM version 6.0, Auckland, New Zealand). The model was interactively fitted to "guide points" provided by the analyst, as well as computer-generated data points calculated from the image using an edge detection algorithm by linear least-squares. The typical time of analysis for a trained expert varied between 24-35 minutes. This finite-element representation enabled a succinct parametrisation with anatomical correspondence across subjects. The spatial representation comprised 215 Bézier parameters ($i = 1 \ldots 215$) which governed the shape of the endocardial and epicardial surfaces [16]. These parameters were expressed in prolate spheroidal coordinates in terms of focal length f (overall scaling) and radial λ_i, hyperboloidal μ_i and azimuthal θ_i coordinates for each control point.

2.1   Temporal Analysis

Functional analysis of the time-varying data [18] comprised two main steps:

1. **Temporal Registration.** Since the number of frames varied with subject, a temporal registration step was needed to ensure that all cases conformed to a common normalised temporal reference (2.1.1).

2. **Temporal Continuity.** A continuous extension through time is desirable for data smoothing, continuity and applying dimension reduction techniques in the time domain. This enables sampling of the models at any time point in the cardiac cycle (2.1.2).

Once these two challenges are overcome, the statistical analysis of linear modes of variation can be written in terms of perturbations about the mean, either for any arbitrary time-point $t = t_i$ in the cardiac cycle (thus becoming static), or by coupling all time variability (discussed in 2.2).

2.1.1   Temporal Registration

To align all cases to a common temporal reference, a time warp from the discrete frame space ($f = 0, 1, 2, \ldots, f_i, \ldots, F$) to a normalised cardiac cycle $[0, 1]$ was constructed such that $t = 0$ represented the ED frame and $t = 0.35$ represented the ES frame (f_{ES}). The normalised time coordinate of 0.35 was chosen for ES because this is the typical normal duration of systole in normal people [7]. This defines a periodic time reference where the warped discrete points $t_i \in [0, 1]$ are given by:

$$t_i = \begin{cases} 0.35 \dfrac{f_i}{f_{ES}} & \text{for } t_i \leq f_{ES} \\[2ex] 0.35 + 0.65 \dfrac{f_i - f_{ES}}{F - f_{ES} + 1} & \text{for } t_i > f_{ES} \end{cases} \tag{1}$$

2.1.2   Temporal Continuity

The Fourier series is a natural representation for our periodic data [2]. Not only does it provide a continuous description of function but it also conveniently represents our function with a small number of coefficients (if we accept some error due to loss of high frequencies).

The Fourier partial sums for any periodic function $f(t)$ at least $\mathcal{L}^1$-integrable in $[-\pi, \pi]$ are

$$(S_N f)(t) = \frac{a_0}{2} + \sum_{n=1}^{N} (a_n \cos(nt) + b_n \sin(nt)) \quad N \geq 0,$$

where $a_n = \dfrac{1}{\pi} \displaystyle\int_{-\pi}^{\pi} f(t) \cos(nt)\, dt\, (n \geq 0)$ and $b_n = \dfrac{1}{\pi} \displaystyle\int_{-\pi}^{\pi} f(t) \sin(nt)\, dt\, (n \geq 1)$.

In our case, the cycle occurs in $[0, 1]$ and we fix the number of harmonics to $N = 5$ which yields 11 coefficients. This has been shown previously to give acceptable error with respect to a high frame rate (60 fps) standard [20].

We therefore have:

$$(S_5 f)(t) = \frac{a_0}{2} + \sum_{n=1}^{5} (a_n \cos(2\pi nt) + b_n \sin(2\pi nt))$$

where $a_n = 2 \displaystyle\int_0^1 f(t) \cos(2\pi t)\, dt\, (n \geq 0)$ and $b_n = 2 \displaystyle\int_0^1 f(t) \sin(2\pi t)\, dt\, (n \geq 1)$.

Given that $f(t)$ must be integrable in $[0, 1]$ and that the available data is discrete with non-uniform spacing (due to the temporal registration), $f(t)$ was chosen to be a cubic B-spline [4] supporting all time-registered points from 2.1.1. This enabled efficient integration by quadrature using QUADPACK [17].

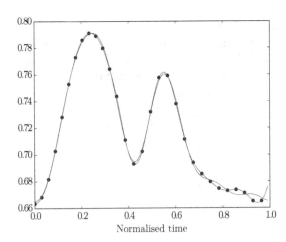

Fig. 1. Example of a $\lambda_i(t)$ parameter. The blue dots represent the time-registered $\lambda_i(t_j)$ points at each frame, the green line the cubic B-spline, and the red line the Fourier partial sums with 5 harmonics $(S_5 f)$.

Figure 1 shows an example of this approximation which leads to two important remarks:

1. By construction, the B-spline function (in green) goes through all available time-points whereas the Fourier approximation $S_5 f$ (in red) can only approximate them since the number of degrees of freedom is smaller than the number of time-points ($11 < 30$ in this particular example)
2. $S_5 f$ is continuously periodic at the boundaries of $[0, 1]$.

Henceforth, for each one of the spatial shape parameters, we use the corresponding $(S_5 \lambda_i)(t)$ as the continuous and smooth temporal extension of our data for statistical time analysis.

2.2   4D Modes of Variation

In order to analyse spatio-temporal variation, two scenarios were built. Let B be the data matrix where the rows represent the different variables and the columns different observations. In our case the variables are the model parameters and the number of observations is $N_1 = 300$ for DETERMINE and $N_2 = 300$ for MESA. Let B_c be a single observation column of B. The first scenario (these results are available on-line[1]) is to simply treat time independently, thus resulting in a variance analysis at a standard sampling of $t \in [0, 1]$, e.g. for ED $B_c^T = [\lambda_1(t_{ED}) \, \lambda_2(t_{ED}) \, \lambda_3(t_{ED}) \, \cdots]$, and for ES $B_c^T = [\lambda_1(t_{ES}) \, \lambda_2(t_{ES}) \, \lambda_3(t_{ES}) \, \cdots]$.

The second scenario, and the one we focus on for the remainder of this paper, is to investigate the spatio-temporal parametric variance. To this end, all 11 Fourier coefficients were coupled into a single vector or column of B. Following the notation in 2.1.2, we then have

$$B_c^T = \left[\underbrace{a_0^{\lambda_1} \, a_1^{\lambda_1} \, b_1^{\lambda_1} \, a_2^{\lambda_1} \, b_2^{\lambda_1} \, \ldots \, a_5^{\lambda_1} \, b_5^{\lambda_1}}_{a_0 + 5 \text{ harmonics}} \quad a_0^{\lambda_2} \, a_1^{\lambda_2} \, b_1^{\lambda_2} \, a_2^{\lambda_2} \, b_2^{\lambda_2} \, \ldots \, a_5^{\lambda_2} \, b_5^{\lambda_2} \, \cdots \right]$$

where for each parameter of the LV model, we have 11 coefficients which carry *most* of the temporal information. This can be interpreted as a multi-variate analysis in shape and time (function) simultaneously, taking advantage of the full physiological information of the finite-element model.

Typically one is only interested in the first few modes of variation, i.e. those which portray most statistical variability. The number of modes that should be kept is a broad topic of research [15] and is dependent on the application.

Figure 2 and Figure 3 show the first three PCA modes of variation for the coupled temporal analysis when using all prolate spheroidal parameters except the focal length (f) for the asymptomatic and symptomatic datasets. To capture 90% of total variation, 22 modes were required for the MESA dataset, whereas the DETERMINE dataset required 27. Temporal animations of these modes and lower-variance modes can be seen on-line[1].

---

[1] http://www.cardiacatlas.org/web/guest/modes

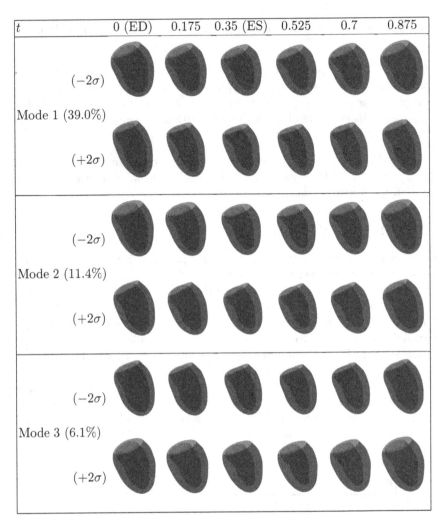

Fig. 2. Asymptomatic (MESA) Fourier temporal modes for all variable prolate spheroidal parameters except focal length (56.5% of variability shown). Slightly elevated anterior view (septum on the left).

3   Discussion

In the MESA or asymptomatic modes of variation in Figure 2, it could be reasoned that the first mode corresponds to the lengthening component of the ventricle, and modes 2 and 3 correspond to features of the mitral valve geometry and base plane tilt. However, from an overall geometric or clinical perspective, there are no *pure* modes of variation, e.g. only sphericity.

In the DETERMINE or symptomatic (infarct) modes in Figure 3, mode 1 represents the sphericity (as was also found in a previous static analysis [13]),

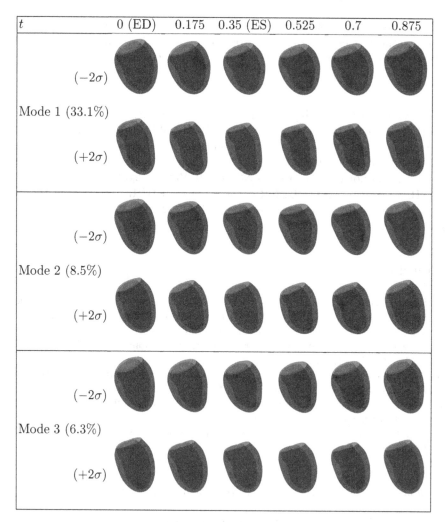

Fig. 3. Symptomatic (DETERMINE) Fourier temporal modes for all variable prolate spheroidal parameters except focal length (47.9% of variability shown). Slightly elevated anterior view (septum on the left).

mode 2 the lower mid-ventricular thickness, mode 3 shows mitral valve geometry features along with a rounding or *bulging* of the apical region. These features correlate well with clinical indicators of heart failure, i.e. sphericity, wall thinning and local dilation of the ventricle are features of infarcted models.

When comparing the modes of variation in Figures 2 and 3 with their static counterparts (available on-line[1]), the first characteristic that becomes apparent is the similarities of the temporal modes with the static counterparts. This implies that the time variability is in itself lesser than the shape variability.

The quantification of these shape and function differences —by projecting onto the atlas modes— enable detection and classification of abnormality by using statistical distances such as Mahalanobis or Bhattacharyya (current ongoing research in our team).

Acknowledgements. The work described was supported by Award Number R01HL087773 from the National Heart, Lung, and Blood Institute (NHLBI). The content is solely the responsibility of the authors and does not necessarily represent the official views of the NHLBI or the National Institutes of Health (NIH). MESA was supported by contracts N01-HC-95159 through N01-HC-95169 from the NHLBI and by grants UL1-RR-024156 and UL1-RR-025005 from the National Center for Research Resources. The NIH (5R01HL091069) and St. Jude Medical provided grant support for the DETERMINE trial.

References

[1] Bild, D., Bluemke, D., Burke, G., Detrano, R., Diez Roux, A., Folsom, A., Greenland, P., et al.: Multi-ethnic study of atherosclerosis: objectives and design. American Journal of Epidemiology 156(9), 871 (2002)

[2] Brown, J., Churchill, R.: Fourier series and boundary value problems. Recherche 67, 2 (1993)

[3] Chandrashekara, R., Rao, A., Sanchez-Ortiz, G.I., Mohiaddin, R.H., Rueckert, D.: Construction of a statistical model for cardiac motion analysis using nonrigid image registration. In: Taylor, C.J., Noble, J.A. (eds.) IPMI 2003. LNCS, vol. 2732, pp. 599–610. Springer, Heidelberg (2003)

[4] Dierckx, P.: Curve and surface fitting with splines. Oxford University Press, USA (1995)

[5] Duchateau, N., De Craene, M., Piella, G., Silva, E., Doltra, A., Sitges, M., Bijnens, B., Frangi, A.: A spatiotemporal statistical atlas of motion for the quantification of abnormal myocardial tissue velocities. Medical Image Analysis 15(3), 316–328 (2011)

[6] Fonseca, C., Backhaus, M., Bluemke, D., Britten, R., Do Chung, J., Cowan, B., Dinov, I., Finn, J., Hunter, P., Kadish, A., et al.: The Cardiac Atlas Project – an imaging database for computational modeling and statistical atlases of the heart. Bioinformatics (2011)

[7] Guyton, A., Hall, J.: Medical Physiology. Saunders, Philadelphia (2000)

[8] Hoogendoorn, C., Duchateau, N., Sánchez-Quintana, D., Whitmarsh, T., Sukno, F., De Craene, M., Lekadir, K., Frangi, A.: A high-resolution atlas and statistical model of the human heart from multislice CT. IEEE Transactions on Medical Imaging (2013)

[9] Kadish, A., Bello, D., Finn, J., Bonow, R., Schaechter, A., Subacius, H., Albert, C., Daubert, J., Fonseca, C., Goldberger, J.: Rationale and design for the defibrillators to reduce risk by magnetic resonance imaging evaluation (DETERMINE) trial. Journal of Cardiovascular Electrophysiology 20(9), 982–987 (2009)

[10] Kaus, M.R., von Berg, J., Niessen, W.J., Pekar, V.: Automated segmentation of the left ventricle in cardiac MRI. In: Ellis, R.E., Peters, T.M. (eds.) MICCAI 2003. LNCS, vol. 2878, pp. 432–439. Springer, Heidelberg (2003)

[11] Lewandowski, A.J., Augustine, D., Lamata, P., Davis, E.F., Lazdam, M., Francis, J., McCormick, K., Wilkinson, A., Singhal, A., Lucas, A., et al.: The preterm heart in adult life: Cardiovascular magnetic resonance reveals distinct differences in left ventricular mass, geometry and function. Circulation (2012)

[12] Lötjönen, J., Kivistö, S., Koikkalainen, J., Smutek, D., Lauerma, K.: Statistical shape model of atria, ventricles and epicardium from short-and long-axis MR images. Medical Image Analysis 8(3), 371–386 (2004)

[13] Medrano-Gracia, P., Cowan, B., Finn, J., Fonseca, C., Kadish, A., Lee, D., Tao, W., Young, A.: The cardiac atlas project: preliminary description of heart shape in patients with myocardial infarction. Statistical Atlases and Computational Models of the Heart, 46–53 (2010)

[14] Medrano-Gracia, P., Cowan, B.R., Bluemke, D.A., Finn, J.P., Lima, J.A.C., Suinesiaputra, A., Young, A.A.: Large scale left ventricular shape atlas using automated model fitting to contours. In: Ourselin, S., Rueckert, D., Smith, N. (eds.) FIMH 2013. LNCS, vol. 7945, pp. 433–441. Springer, Heidelberg (2013)

[15] Mei, L., Figl, M., Darzi, A., Rueckert, D., Edwards, P.: Sample sufficiency and PCA dimension for statistical shape models. In: Forsyth, D., Torr, P., Zisserman, A. (eds.) ECCV 2008, Part IV. LNCS, vol. 5305, pp. 492–503. Springer, Heidelberg (2008)

[16] Nielsen, P., Le Grice, I., Smaill, B., Hunter, P.: Mathematical model of geometry and fibrous structure of the heart. American Journal of Physiology- Heart and Circulatory Physiology 260(4), H1365 (1991)

[17] Piessens, R., Doncker-Kapenga, D., Überhuber, C., Kahaner, D., et al.: QUADPACK, a subroutine package for automatic integration. Springer (1983)

[18] Ramsay, J.: Functional data analysis. Wiley Online Library (2006)

[19] Young, A., Cowan, B., Thrupp, S., Hedley, W., Dell'Italia, L.: Left ventricular mass and volume: Fast calculation with guide-point modeling on MR images. Radiology 216(2), 597 (2000)

[20] Young, A.A., Hunter, P.J., Smaill, B.H.: Estimation of epicardial strain using the motions of coronary bifurcations in biplane cineangiography. IEEE Transactions on Biomedical Engineering 39(5), 526–531 (1992)

Progress on Customization of Predictive MRI-Based Macroscopic Models from Experimental Data

Mihaela Pop[1,2,*], Maxime Sermesant[3], Samuel Oduneye[1,2], Sudip Ghate[1],
Labonny Biswas[1], Roey Flor[1], Susan Newbigging[2,4], Eugene Crystal[1,2],
Nicholas Ayache[3], and Graham A. Wright[1,2]

[1] Sunnybrook Research Institute, Toronto
mihaela.pop@utoronto.ca
[2] University of Toronto, Canada
[3] Inria - Asclepios Project,
Sophia Antipolis, France
[4] CMHD Pathology Core, Toronto, Canada

Abstract. MR image-based computer heart models are powerful non-invasive tools that can help us predict the transmural electrical propagation of abnormal depolarization-repolarization waves in the presence of infarct scars (i.e., collagenous fibrosis), a major cause of sudden death; however, an important step is the customization of these models from electrophysiology studies (EP). In this work, we used MR-EP data obtained in a pre-clinical animal model (i.e., three healthy and two infarcted swine hearts) and customized a simple mono-domain model (i.e., the Aliev-Panfilov model). Specifically, we estimated the mathematical parameters corresponding to: a) the repolarization phase from *in vivo* activation-recovery intervals, ARIs (recorded *in vivo* with a CARTO system), and b) the anisotropy ratio (from fluorescence microscopic imaging of connexin 43, Cx43). Our measurements showed that in the ischemic peri-infarct areas the ARIs intervals were shorter by ~ 14% compared to those in normal tissue, and that there was a significant reduction (> 50%) in the Cx43 density (which tunes the cell-to-cell coupling and tissue bulk conductivity) with respect to both longitudinal and transverse directions of the myocyte. In addition, we included comparisons between virtual *in silico* simulations of activation maps obtained with different parameters used as input to a 3D MR-based biventricular model. Our preliminary results demonstrated the feasibility of using *generic* parameters to customize such MR-based models; however, further quantitative studies are needed. Finally, we discussed the overall advantages and limitations of our simplified approach, along with future directions.

Keywords: cardiac MRI, modelling, electrophysiology, histopathology.

---

* Corresponding author.

O. Camara et al. (Eds.): STACOM 2013, LNCS 8330, pp. 152–161, 2014.

1   Introduction

Abnormal propagation of the electrical wave in patients with structural disease (e.g. myocardial infarct) is a major cause (> 85%) of sudden cardiac death due to lethal ventricular arrhythmias (such as ventricular tachycardia, VT, and ventricular fibrillation, VF) [1]. An important task is the evaluation of chronic fibrosis in post-infarction, together with the quantification of structural and electrical properties changes due to the scar and peri-infarct (i.e., the VT substrate, which is the target of RF ablation) [2, 3]. To achieve this, imaging and electrophysiology (EP) methods as well as computational tools have been continuously refined [4, 5, 6]. Integration of myocardial electrical and structural characteristics, along with the model customization from measurements, are complex processes that include experimentation using clinical EP tools, non-invasive imaging and modelling [7, 8].

Our broad aim is to predict the propagation of the electrical wave using MRI-based computer models enriched with *in vivo* EP measurements obtained in healthy and chronically infarcted swine. Such a pre-clinical model is advantageous to use in a translational experimental-modelling framework, because the swine heart size is close to the human heart. Previously, we focused on: a) building 3D MRI-based models from high resolution diffusion-weighted MR imaging (which also enabled the incorporation of fiber directions); and b) customizing the model parameter corresponding to a *global* 'bulk' conductivity of tissue [9, 10]. For the latter, the customization step was performed by calculating the speed of depolarization wave from the local activation times, LAT, measured at precise locations (i.e., determined from the spatial coordinates of the catheter tip).

In this current work, we report further progress within our experimental-modelling framework and we focus on tuning other important parameters in the computer model. For instance, we customized the parameter corresponding to the 'recovery' phase from measured ARIs, a clinical surrogate of the action potential duration, APD [11]. Furthermore, we also derived the anisotropy ratio for tissue conductivity from fluorescence microscopy imaging of connexin 43 (Cx43) protein, a major ventricular gap junction that facilitates the flow of the ionic current between the cells, and therefore tunes the electrical conductivity [1]. Typically, in healthy myocardium, Cx43 proteins have a higher abundance in the intercalated disks (connecting the cells edge-to-edge), resulting in two-three times faster propagation of the activation wave in the direction parallel to the myocytes; however, this ratio changes in ischemic conditions due to the gap junctions closure. These measurements (i.e., ARI and Cx43 density) were used to determine *generic* parameters in the Aliev-Panfilov model [12] and to predict *in silico* the wave propagation using a 3D MR-based heart model previously developed (with fiber directions integrated from diffusion-weighting, DW) [10]. Figure 1 shows a simplified diagram of the workflow in the experimental-theoretical framework.

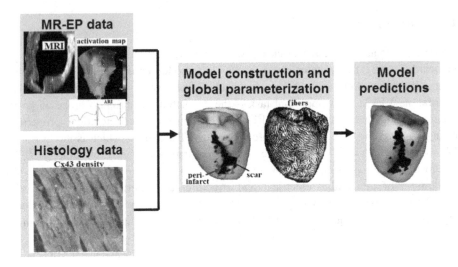

Fig. 1. The integration of experimental data into predictive MRI-based computer models

2   Material and Methods

2.1   The Experimental Data

In this paper we included results from five EP studies performed in a swine pre-clinical model (i.e., three healthy and two chronically infarcted hearts, at 5 weeks post-infarction), approved by our Sunnybrook Research Institute (Toronto, Canada). The methodology of generating infarctions as well as the procedures steps associated with the EP studies in healthy and infarcted hearts was previously described [9, 10]. Briefly, the activation maps were recorded using an invasive contact electro-anatomical mapping system (i.e., CARTO-XP, Biosense, USA), using filters applied to measure only signals within 30-400Hz. Here, ARIs, were determined for all intra-cardiac unipolar waves using a CARTO analysis software, as explained in [11]. For each wave, a local repolarization time (LRT) was found for biphasic and negative deflections using dV/dt_{max} (maximum rate of rise of voltage), whereas for positive deflections we used dV/dt_{min} on the descending limb of the T-wave. Finally, ARI was calculated as the difference between LRT and LAT. All calculations and annotations were manually performed off-line.

In addition, samples from the two infarcted hearts were taken from the healthy myocardium, peri-infarct and dense scar. The samples were fixed in 10% formalin and embedded in paraffin. Thin slices (4-5μm) were fixed and stained with Picrosirius Red for collagen assessment. Adjacent slices were also stained and prepared for fluorescence microscopic imaging of Cx43 as in [13]. The Cx43 density was quantified from fluorescence images on select ROIs. We assessed the alteration in cell-to-cell coupling (longitudinal direction) and side-to-side coupling (transverse direction) using the *Visiopharm* software tool (www.visiopharm.com).

2.2   Mathematical Formalism and Computer Model

The **Aliev-Panfilov (A-P) Model** is based on reaction-diffusion type of equations and solves for the action potential (V) and recovery term (r) as described in [12]:

$$\frac{\partial V}{\partial t} = \nabla \cdot (D\nabla V) - kV(V-a)(V-1) - rV \tag{1}$$

$$\frac{\partial r}{\partial t} = -(\varepsilon + \frac{\mu_1 r}{\mu_2 + V})(kV(V-a-1) + r) \tag{2}$$

One parameter of interest is a, which tunes the action potential duration, APD. Other parameters corresponding to recovery phase ($k=8$, and $\mu1\&\mu2$) were given in [14]. This simplified model accounts for the heart muscle structural anisotropy (i.e., fiber directions) via the diffusion tensor D, which can be written as:

$$D = d\begin{pmatrix} 1 & 0 & 0 \\ 0 & \rho & 0 \\ 0 & 0 & \rho \end{pmatrix}$$

where d is the 'bulk' conductivity of tissue. The anisotropy ratio ρ depends on the conductivities in the transverse and longitudinal direction of myocytes. At a cellular level, this ratio is generated by a heterogeneous distribution of connexin Cx43 in the intercalated disks vs. on the surrounding sheath [1]. In ischemia, ρ is altered [1], changing the solution to the action potential in eq. (2). Furthermore, a reduced value of d results in a slower propagation of activation wave, as per the relation between the propagation speed c and 'bulk' conductivity d [7]:

$$c = \sqrt{2 \cdot k \cdot d}(0.5 - a) \tag{3}$$

3   Results

3.1   Estimation of Model Parameters from Experimental Data

Figure 2 shows an exemplary unipolar CARTO wave recorded in a healthy swine in sinus rhythm, along with the interpolated ARI map on the LV-endocardium (the white dots correspond to the valve). Approximately 470 points were recorded in the hearts (n=5) included in this paper. We obtained a mean ARI value of ~306±11ms for healthy zones and ~265±19ms for peri-infarct zones, and these ARI values were considered surrogates for APD90 corresponding to these two myocardial zones.

Figure 3 shows a theoretical calibration curve a vs. $APD90$ generated with the Aliev-Panfilov mathematical model, from which we extracted the a values of interest. Specifically, by using fitting functions implemented in Matlab (Mathworks) as in [12] and [15], the mean values for measured ARIs yielded the following values for a: 0.113 for healthy zones and 0.116 for peri-infarct, respectively. Note that here k was kept fixed, as we only derived generic a values for sinus rhythm cases (i.e., heart cycle length usually longer than 600 ms).

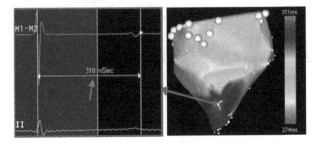

Fig. 2. Determination of ARI from the unipolar waves recorded by the CARTO system

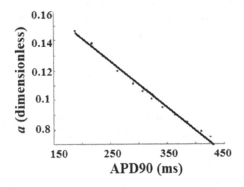

Fig. 3. Theoretical calibration curve used to derive generically the parameter "*a*"

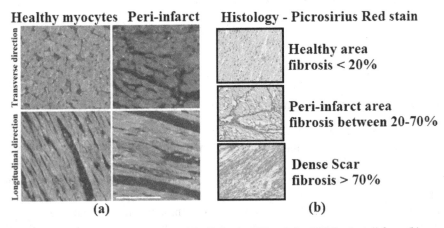

Fig. 4. (a) Fluorescence micrographs of Cx43 (scale 100 μm) for ROIS selected from (b) areas with different fibrosis severity in Picrosirius Red stain

Figure 4 shows histopathology results obtained from samples containing healthy, peri-infarct and dense scar. A quantitative analysis demonstrated a significant reduction (> 50%) in Cx43 density in the peri-infarct as compared to the healthy myocardium; specifically, we measured a ~61% reduction of Cx43 in the longitudinal direction and ~52% in the transverse direction. This reduction can be qualitatively observed in the light micrographs of fluorescence Cx43 included in Fig 4a. These values correspondingly lead to reduced conduction velocities in the normal and peri-infarct areas, as per eq. (3). The analysis from selected ROIs, also yielded the following conductivity ratios (transverse : longitudinal) $(1:2.53)^2$ for the normal myocardium and $(1:2.07)^2$ for the peri-infarct areas, respectively.

Notably, the myocardial tissue was categorized by an expert based on our novel grading system for fibrosis [18] into: healthy, peri-infarct and scar, using the Picrosirius Red stain, where collagen stains red and the healthy myocytes stain yellow (see Fig 4b).

3.2   Impact of Model Parameters on Simulation Results

To illustrate the impact of these *generic* parameters on activation maps, we included below several results from *in silico* simulations obtained using a 3D MRI-based biventricular heart model, previously constructed [10]. For simplicity, we virtually paced the heart from the apex of RV-endocardium, and observed the propagation of action potential under different combinations of model parameters. Figure 5a shows the 3D model with the pacing site and the scar (in the LCX territory) indicated by arrows, whereas Fig 5b shows the resulting simulated action potential waves in the

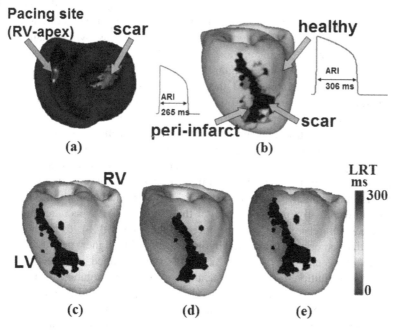

Fig. 5. Simulation results obtained for an *in silico* 3D MR image-based model (a-b) using different input model parameters (c-d-e) (see text for details)

healthy and peri-infarct areas. The *a* values determined in Section 3.1 were set such that the APD90 values were equal to the mean ARI values obtained in the CARTO measurements. Figures 5c, 5d and 5e represent repolarization maps. These were obtained using different combinations of model parameters, where a 'bulk' conductivity in the normal tissue was set to $d=3$ to tune a conduction velocity of ~80 cm/s in the direction parallel to the myocytes axes.

The other model parameters were set to obtain the following combinations: (5c) isotropic healthy and peri-infarct tissue, (5d) reduced conductivity *d* in the peri-infarct by 50% and the same anisotropic conduction velocity ratio (1:2) in both healthy and peri-infarct zones, and (5e) a model customization per zones with the ratio of conduction velocities as obtained in Section 3.1.

4   Discussion and Future Work

Non-invasive evaluation methods like cardiac MR imaging and predictive image-based computer models are becoming powerful tools for the clinician, as they can supplement the surfacic invasive EP data by providing transmural information such as: infarct location/extent and 3D propagation of the electrical wave through the heart. The integration of electrical and structural information will help us target such predictive models to desired clinical applications. However, the alteration of myocardial structure/function and corresponding tissue heterogeneities need to be quantified and integrated into appropriate predictive models. Hence, the estimation of generic parameters, the customization of image-based computer models, as well as the validation/testing under controlled experimental conditions, will remain important prerequisites prior to the integration of such MR image-based cardiac computer models into the clinical treatment planning platforms.

In this work, we demonstrated the feasibility of estimating the Aliev-Panfilov model parameters corresponding to the repolarization phase and the anisotropy ratio. First, our experimental repolarization values (i.e., shorter ARI in the peri-infarct compared to the healthy myocardium) are consistent with the results obtained in a previous *ex vivo* optical study [15], and appear to be a hallmark of the changes in electrical properties at ~5-6 weeks after infarction. It is well known that in the acute phase (hours after insult) the APD becomes dramatically shorter due to hypoxia and altered pH, after which, in the sub-acute & early chronic phases, the APD starts to recover and eventually, month-years after the insult, the APD becomes longer in the ischemic peri-infarct in other large hearts (dog and human) [16]. A limitation is that, so far, we derived only *a* from sinus rhythm measures. At higher (pacing) frequencies, another recovery parameter (i.e., *k*) needs to be determined in order to reproduce correctly the restitution curve; however, there is no unique solution for the (*a,k*) pair (a disadvantage of the Aliev-Panfilov model). Lastly, we should mention that for the Aliev-Panfilov mono-domain model, the simulation time of 0.8s of the heart cycle on a mesh of approx. 200,000 elements (element size of ~1.2 mm), was about 40 min on an Intel ® Core™ 2 duo CPU, T5550 @1.83GHz, with 4 GB of RAM. With this respect, the Mitchell-Schaeffer physiological model might be more accurate to use and much faster if implemented using GPU [19].

We acknowledge that we used a simplified anisotropic model, which is distinct from the orthotropic models in that it only includes the fiber directions and neglects the sheet-like anisotropy and correspondingly, the (more realistic) three conductivity tensors. Although limited to measurements taken relative the fiber directions (and not including the sheets), our results showed that ρ was altered in the ischemic peri-infarct area, where Cx43 density was reduced by > 50% in both directions (longitudinal and transverse, respectively), consistent with other studies [17]. The redistribution of Cx43 triggers a reduced conduction velocity along with abnormal activation patterns as revealed by other studies [1], with conduction velocity considerably reduced in the peri-infarct areas. Such Cx43 measurements offered an alternative to the extraction of ρ from surfacic EP measurements (using the CARTO system) because the conduction velocity estimated from the depolarization times depends on the density of the recorded points (which are sparse around the peri-infarct and usually only from the endocardium).

Overall, this study underlies the feasibility of customizing predictive MR image-based models with generic model parameters derived from experimental data. In the myocardial infarction case, these model parameters should reflect the structural and electrical properties corresponding to a specific healing phase. Our parameters have to be used cautiously, as they are derived from a small set of data; thus, more experiments are needed to confirm these values as well as a quantitative comparison between the activation times predicted by the computer model and the measured depolarization-repolarization maps. From the experimental point of view, we need to reduce: a) the total procedure time associated with a typical *in vivo* MR study followed by an EP study, and b) the errors between the location of scar & peri-infarct identified by MR imaging and by the EP system. Here, an alternative would be to perform the EP study under real-time MR guidance. Such system has been successfully implemented and feasibility studies yielded promising results [20], with the MR-guided EP mapping producing a significantly lower average location error (i.e., a point to surface distance mean error of 2.1±1.1 mm when compared to a 4.8±2.0 mm error obtained using CARTO). We illustrate below an example from a real-time EP-MR study performed in our laboratory, where the data was visualized using the open-source Vurtigo tool (www.vurtigo.ca). Figure 6 (*left panel*) shows the MR-compatible EP catheter maneuvered inside the LV cavity, together with a myocardial tissue classification from contrast-enhanced MR images in an infarcted swine. In the *right panel,* we included an interpolated depolarization map reconstructed over the LV endocardial surface, together with examples from two electrical signals (waves) from the healthy myocardium (ARI ~ 300ms) and ischemic peri-infarct (ARI ~ 270ms), which are within the range of the CARTO-based values derived in this paper and are also close to those derived by us from optical imaging, *ex vivo* [15]. These preliminary results, together with the lessons learned from the experimental-modelling framework integrating MR imaging and CARTO studies, give us confidence to anticipate successful implementation of predictive models using the real-time MR-guided EP studies and potentially integrate fast computer models into such platforms [19].

Real time MRI-guided EP study

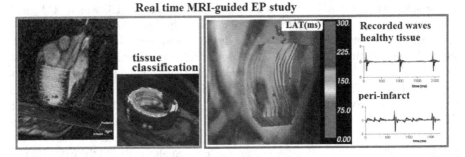

Fig. 6. Example of a real-time MRI-EP study (see text for details)

To conclude, computational models should include sufficient level of details at both anatomical and electrophysiological levels, and should capture accurately the structure and electrical function of the myocardium (in healthy and ischemic states). To do so, collaborative and multi-disciplinary efforts must be at the center of future research, and this will pave the success of translating the pre-clinical models into clinical platforms for image-guidance of cardiac EP procedures.

Acknowledgement. This work was financially supported in part by a CIHR grant (MOP93531). We thank Mrs. Jen Barry for help with the animal experiments, Dr. Tommaso Mansi for help with the mesh generation, and Ms Lily Morikawa for help with the histopathological processing and analysis of the tissue samples.

References

1. Janse, M.J., Wit, A.L.: Electrophysiological mechanisms of ventricular arrhythmias resulting from myocardial ischemia and infarction. Physiol. Rev. 69(4), 1049–1169 (1989)
2. Stevenson, W.G.: Ventricular scars and VT tachycardia. Trans. Am. Clin. Assoc. 120, 403–412 (2009)
3. Bello, D., Fieno, D.S., Kim, R.J., et al.: Infarct morphology identifies patients with substrate for sustained ventricular tachycardia. J. Am. College of Cardiology 45(7), 1104–1110 (2005)
4. Codreanu, A., Odille, F., Aliot, E., et al.: Electro-anatomic characterization of post-infarct scars comparison with 3D myocardial scar reconstruction based on MR imaging. J. Am. Coll. Cardiol. 52, 839–842 (2008)
5. Clayton, R.H., Panfilov, A.V.: A guide to modelling cardiac electrical activity in anatomically detailed ventricles. Progress in Biophysics & Molecular Biology – Review 96(1-3), 19–43 (2008)
6. Detsky, J.S., Paul, G., Dick, A.J., Wright, G.A.: Reproducible classification of infarct heterogeneity using fuzzy clustering on multi-contrast delayed enhancement magnetic resonance images. IEEE Trans. Med. Imaging 28(10), 1606–1614 (2009)
7. Chinchapatnam, P., Rhode, K.S., Ginks, M., Rinaldi, C.A., Lambiase, P., Razavi, R., Arridge, S., Sermesant, M.: Model-Based imaging of cardiac apparent conductivity and local conduction velocity for planning of therapy. IEEE Trans. Med. Imaging 27(11), 1631–1642 (2008)

8. Sermesant, M., Delingette, H., Ayache, N.: An electromechanical model of the heart for image analysis and simulations. IEEE Transactions in Medical Imaging 25(5), 612–625 (2006)
9. Pop, M., Sermesant, M., Flor, R., Pierre, C., Mansi, T., Oduneye, S., Barry, J., Coudiere, Y., Crystal, E., Ayache, N., Wright, G.A.: *In vivo* Contact EP Data and *ex vivo* MR-Based Computer Models: Registration and Model-Dependent Errors. In: Camara, O., Mansi, T., Pop, M., Rhode, K., Sermesant, M., Young, A. (eds.) STACOM 2012. LNCS, vol. 7746, pp. 364–374. Springer, Heidelberg (2013)
10. Pop, M., Sermesant, M., Mansi, T., Crystal, E., Ghate, S., Peyrat, J.M., Lashevsky, I., Qiang, B., McVeigh, E.R., Ayache, N., Wright, G.A.: Correspondence between simple 3D MRI-based computer models and in-vivo EP measures in swine with chronic infarctions. IEEE Transactions on Biomedical Engineering 58(12), 3483–3486 (2011)
11. Gepstein, L., Hayam, G., Ben-HAim, S.A.: Activation-repolarization coupling in the normal swine endocardium. Circulation 96, 4036–4043 (1997)
12. Aliev, R., Panfilov, A.V.: A simple two variables model of cardiac excitation. Chaos, Soliton and Fractals 7(3), 293–301 (1996)
13. Zhang, Y., Wang, H., Kovacs, A., Kanter, E.M., Yamada, K.A.: Reduced expression of Cx43 attenuates ventricular remodelling after myocardial infarction via impaired TBF-B signaling. American Journal of Physiology, Heart and Circ. Physiol 298(2), H477–H487 (2010)
14. Nash, M.P., Panfilov, A.V.: Electromechanical model of excitable tissue to study reentrant cardiac arrhythmias. Prog. Biophys. Molec. Biol. 85, 501–522 (2004)
15. Pop, M., Sermesant, M., Liu, G., Relan, J., Mansi, T., Soong, A., Peyrat, J.-M., Truong, M.V., Fefer, P., McVeigh, E.R., Delingette, H., Dick, A.J., Ayache, N., Wright, G.A.: Construction of 3D MR image-based computer models of pathologic hearts, augmented with histology and optical imaging to characterize the action potential propagation. Medical Image Analysis 16(2), 505–523 (2012)
16. Ursell, P.C., Gardner, P.I., Albala, A., Fenoglio, J., Wit, A.L.: Structural and electrophysiological changes in the epicardial border zone of canine myocardial infarcts during infarct healing. Circulation Res. 56, 436–451 (1985)
17. Jansen, J., van Veen, T.A.B., de Jong, S., vand der Nagel, R., van Rijen, H.V.M., et al.: Reduced Cx43 expression triggers increased fibrosis due to enhanced fibroblast activity. Circulation Arrhythmia and Electrophsiology 5, 380–390 (2012)
18. Pop, M., Ghugre, N., Ramanan, V., Morikawa, L., Stanisz, G., Dick, A.J., Wright, G.A.: Quantification of fibrosis in infarcted swine hearts by ex vivo late gadolinium enhancement and diffusion-weighted MRI methods. Physics in Medicine and Biology 58(15), 5009–5028 (2013)
19. Talbot, H., Duriez, C., Courtecuisse, H., Relan, J., Sermesant, M., Cotin, S., Delingette, H.: Towards Real-Time Computation of Cardiac Electrophysiology for Training Simulator. In: Camara, O., Mansi, T., Pop, M., Rhode, K., Sermesant, M., Young, A. (eds.) STACOM 2012. LNCS, vol. 7746, pp. 298–306. Springer, Heidelberg (2013)
20. Oduneye, S.O., Biswas, L., Ghate, S., Ramanan, V., Barry, J., Laish-FarKash, A., Kadmon, E., Zeidan Shwiri, T., Crystal, E., Wright, G.A.: The feasibility of endocardial propagation mapping using MR guidance in a swine model and comparison with standard electro-anatomical mapping. IEEE Trans. Med. Imaging 31(4), 977–983 (2012)

Automatic Personalization of the Mitral Valve Biomechanical Model Based on 4D Transesophageal Echocardiography

Jingjing Kanik[1,*], Tommaso Mansi[4], Ingmar Voigt[4], Puneet Sharma[4],
Razvan Ioan Ionasec[4], Dorin Comaniciu[4] and James Duncan[1,2,3,*]

[1] Department of Biomedical Engineering, Yale University, New Haven, CT, USA
[2] Electrical Engineering, Yale University, New Haven, CT, USA
[3] Diagnostic Radiology, Yale University, New Haven, CT, USA
{jingjing.zhu,james.duncan}@yale.edu
[4] Imaging and Computer Vision, Siemens Corporation, Corporate Technology,
Princeton, NJ, USA

Abstract. Patient-specific computational models including morphological and biomechanical models based on medical images have been proposed to provide quantitative information to aid clinicians for Mitral Valve (MV) disease management. Morphological models focus on extracting geometric information by automatically detecting the mitral valve structure and tracking its motion from medical images. Biomechanical models are primarily used for analyzing the underlying mechanisms of the observed motion pattern. The recently developed patient-specific biomechanical models have integrated the personalized mitral apparatus and boundary conditions estimated from medical images to predicatively study the pathological changes and conduct surgical simulations. As a next step towards transitioning patient-specific models into clinical settings, an automatic personalization algorithm is proposed here for biomechanical models extracted from Transesophageal Echocardiography (TEE). The algorithm achieves the customization by adjusting both the chordae rest length and material parameters such as Young's modulus which are challenging to estimate or measure directly from the medical images. The algorithm first estimates the mitral valve motion from TEE using a machine learning method and then fits the biomechanical model generated motion into the image-based estimation by minimizing the Euclidean distances between the two. The algorithm is evaluated on 4D TEE images of five patients and yields promising results, with an average fitting error of $1.84 \pm 1.17mm$.

1 Introduction

Medical imaging techniques provide powerful tools to visualize valvular structures with echocardiography being the most widely used modality in clinical applications because of its high temporal resolution, ease of use and relatively

---

* Corresponding author.

O. Camara et al. (Eds.): STACOM 2013, LNCS 8330, pp. 162–170, 2014.

low cost. The newly developed 4D Transesophageal Echocardiography (TEE) acquires clearer images with higher temporal resolution thus is now the prefered modality for valve assessment. The advancement in imaging techniques allows for more accurate quantitative evaluation and modeling of the Mitral Valve (MV) structure. Quantitative patient-specific modeling tools are demanded to aid predictive surgical planing to achieve optimal treatment in clinical practice. [1]

Several approaches have been proposed to model MV geometry and dynamics, including morphological and biomechanical models. The morphological models employ an automatic or semi-automatic method to detect the mitral apparatus and track its motion from medical images[3,4,5,6]. These models can provide visualization and quantitative measurements of the anotomical structure, but do not explain the underlying mechanisms of the motion pattern or pathological changes. Several patient-specific biomechanical models [2,7] of MV have been proposed using geometric information from medical images and general material parameters of the mitral leaflet tissues from experimental results. Mansi et al. [2] proposed a patient-specific biomechanical model and simulated MitralClip intervention with comparison to real outcome. Such models have great potential to become efficient predictive tools to design preoperative treatment plans in selecting the patients and determining clipping sites to ensure the optimal outcome. Votta et al. [1] reviewed recent advancement in computational biomechanical models and identified areas of improvement to develop clinically applicable tools. Realistic morphological and functional information is essential for a comprehensive patient-specific model.

To facilitate the transition of patient-specific model to clinical applications, a user-friendly platform has to be built for the clinicians to easily set up the model for their purpose without going through complex training. An automatic personalization algorithm is proposed in this paper to customize the biomechanical model based on TEE images without any user interaction. The algorithm automatically detects mitral valve motion from TEE images and fits the biomechanical model into the observed motion following a two step procedure. The first step estimates the chordae rest length which can have a significant influence on mitral valve dynamics[2], using a coarse-to-fine maximum derivative approach. The second step estimates the material parameters such as Young's modulus, using Extended Kalman Filter (EKF)[8,9]. The algorithm provides a framework for personalization of the biomechanical model where additional parameters of interest can also be estimated in more complex models. The algorithm is evaluated on TEE images of five patients with promising results.

2   Method

2.1   Overview of the Algorithm

In this paper, we study the mitral valve closure process from the end diastole, which is the last frame (I_0) where the mitral valve is seen fully open in TEE images, to the first systolic frame (I_N) where the mitral valve is seen maximally closed. The algorithm first estimates the leaflet geometry at the nth frame (g_n)

and tracks its motion to ensure inter- and intra- patient point correspondence of the geometric representation. The biomechanical model generated motion sequence $(h(g_0, m))$ is then fit into the image-based observation by adjusting a set of patient-specific parameters (m), which are composed of leaflet biomechanical parameters and the chordae rest length. The estimation problem is formulated as follows:

$$m = \min_m f(m) = \min_m \|g_n - h(g_0, m)\| \qquad (1)$$

where the cost function is represented by the Euclidean distances between biomechanical model generated and image observed mitral valve closure. Model personalization is achieved by minimizing the cost function to obtain the patient-specific parameters. The cost function can be modified to penalize the mismatch in degree of coaptation for certain clinical applications when matching at the leaflet edge is more important than other regions. To solve the optimization problem, a two-step procedure is followed as illustrated in Fig 1.

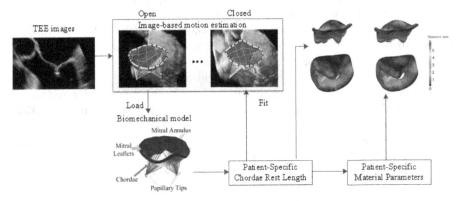

Fig. 1. Overview of the algorithm

2.2   Morphological Model

The mitral valve apparatus, including the mitral annulus, anterior and posterior leaflets, and papillary tips, is extracted from end diastole and early systole using the hierarchical discriminative learning algorithm described in [5]. The inter- and intra- patient point correspondence is achieved through a manifold-based motion model. The mitral leaflets are represented by triangulated surface meshes as shown in Fig. 1. The geometry at the end diastole is then further processed to be loaded into the biomechanical model. First, the one layer leaflet surface mesh is extruded towards the ventricle for a set distance, which is 1.32 mm and 1.26 mm for the anterior and posterior leaflet respectively, to move the layer closer to the left ventricle thus forming a volumetric structure. Next, the volumetric structure is discretized into tetrahedral meshes. This is followed by mapping fiber models on the leaflets as in [2], where the fiber directions are mainly parallel to the annulus while those in the anterior leaflet close to the commissures gradually rotate to become perpendicular to the annulus. Finally, the chordae are attached between the leaflet and the papillary tips including twenty eight marginal chordae and eight basal chordae. The insertion points are determined by visual inspection and are identical for all the patients. The geometry at early systole are processed in a similar manner and used in the automatic personalization process.

2.3 Biomechanical Model

The mitral leaflets are modeled as linear, transversely isotropic and nearly incompressible elastic tissues [2]. The tissue material properties, including Young's modulus along and across the collagen fiber and shear modulus (E_{AL_f}, $E_{AL_{f\perp}}$, G_{AL}, E_{PL_f}, $E_{PL_{f\perp}}$ and G_{PL} respectively), of the anterior and posterior leaflet are assumed to be different for different patients. The Poisson's ratio (ν) is set to be 0.488 to capture the incompressible nature of the tissues for both leaflets. The mitral valve dynamics is simulated using a finite element method to solve the dynamic equation:

$$M\ddot{U} + C\dot{U} + KU = F_c + F_p \qquad (2)$$

where U, $\dot{U}$ and $\ddot{U}$ are the displacement, velocity and acceleration vectors of the vertices of the mitral valve mesh respectively. M is the diagonal mass matrix (a uniform mass density $\rho = 1.04g/ml$ is used), K is the stiffness matrix and a function of material parameters, and C is the Rayleigh damping matrix (C=0.1(M+K)). F_p is the force developed by the heart pressure and modeled by a generic pressure profile that increases from 0mmHg to 120mmHg. F_c is the force induced by the chordae which is calculated using the following equation:

$$F_c(v_i, p_i, t) = -k_{c,i}(\epsilon_{c,i}, t) \times (L_i(t) - L_{i,0}) \qquad (3)$$

where $L_i(t)$ is the current elongation, $L_{i,0}$ is the chordae rest length, $\epsilon_{c,i}(t) = (L_i(t) - L_{i,0})/L_{i,0}$ is the strain, $k_{c,i}$ is the spring tensile stiffness and related to chordae material properties. The mitral annulus motion and the papillary tip motion derived from the TEE images are used as the boundary condition for the biomechanical model. The biomechanical model is implemented in SOFA[1] . The target set of patient-specific parameters are defined as

$$\mathbf{m} = [E_{AL_f}, E_{AL_{f\perp}}, G_{AL}, E_{PL_f}, E_{PL_{f\perp}}, G_{PL}, L_{1MA}, \ldots, L_{14MA},$$
$$L_{1MP}, \ldots, L_{14MP}, L_{1BA}, \ldots, L_{4BA}, L_{1BP}, \ldots, L_{4BP}]$$

where $L_{MA}, L_{MP}, L_{BA}, L_{BP}$ are the chordae rest length of the marginal and basal chordae attached to anterior and posterior papillary tips.

2.4 Personalization of the Biomechanical Model

The goal of personalization is to determine a set of parameters that minimizes the distance (f(m)) between the biomechanical model driven and the image-observed mitral valve closure. The first step of the algorithm aims to personalize the rest length using a coarse-to-fine maximum derivative method

---

[1] http://www.sofa-framework.org/

Algorithm 1. Coarse-to-fine maximum derivative

1. Initialize the chordae rest length using the point-to-point distance from the papillary tip and the insertion points at the end systole
2. At Jth level, change the group of the parameters in the direction of maximum derivative to reduce the cost function
3. Repeat 2 until the cost function does not change between two consecutive iterations
4. Go to the (J+1)th level and repeat 2,3

as shown in Algorithm 1. Twenty-eight marginal chordae are used, fourteen of which attached to each leaflet, and seven attached to each papillary tip. Eight basal chordae are used, four of which are attached to each leaflet, and two attached to each papillary tip. Fixing the material parameters, thirty-six parameters need to be estimated in the first step ($\mathbf{m_1} = [L_{1MA}, \dots, L_{14MA}, L_{1MP},$ $\dots, L_{14MP}, L_{1BA}, \dots, L_{4BA}, L_{1BP}, \dots, L_{4BP}]$). The estimation starts from the first and coarsest level which has eight groups of parameters and ends at the finest level where each chordae rest length is estimated individually. Grouping is determined by their location where seven marginal chordae form one group while two basal chordae form the other group. The grouping become finer at each level and the number of chordae included in each group is halved until each chordae rest length is estimated individually. Seven groups of marginal chordae and eight groups of basal chordae are used in the second level. Fourteen groups of maginal chordae are used in the third level and each of the marginal chordae rest length are estimated individually in the fourth level. The method provides better computational efficiency since the optimization at the coarse level provides a better starting point for finer tuning.

The second step of the algorithm aims to personalize material parameters using an Extend Kalman filter (EKF) approach since EKF provides the stable sequential least square solution and has been shown to be efficient for material parameter estimation in [8,9]. Once the chordae rest length is fixed, there are six parameters ($\mathbf{m_2} = [E_{ALf}, E_{ALf\perp}, G_{AL}, E_{PLf}, E_{PLf\perp}]$) to be estimated, four of which can be derived from the other two. The ratio of Young modulus along and across the fiber ($r = E_f/E_{f\perp}$) is fixed and the shear modulus is approximated by $G \approx E_f/(2((1+\nu)))$ to ensure the physiological consistency of the parameters. The state space representation is written as follows:

$$\mathbf{m_{2,k}} = f(\mathbf{m_{2,k-1}}) + w_{k-1} = \mathbf{m_{2,k-1}} + w_{k-1}$$

$$\mathbf{g_k} = h(\mathbf{m_{2,k}}) + v_k$$

where w_{k-1} and v_k are the state and process noises respectively and assumed to follow Gaussian distributions with covariance matrix Q_k and R_k. The observation vector $\mathbf{g_k} = [x_{k1}, y_{k1}, z_{k1}, \dots, x_{ki}, y_{ki}, z_{ki}, \dots, x_{kL}, y_{kL}, z_{kL}]$ is the geometry vector which is represented by L number of vertices (L=3248 in this study).The process function $f(.)$ is derived from the assumption that material parameters and the chordae rest length stay constant during the cardiac cycle. The observation function $h(.)$ is derived from the biomechanical model specifying loading, geometry, tissue property, boundary condition, and dynamic equilibrium function and is the same as in the cost function (eq. 1).

The EKF estimation is first initialized with the general material parameters $(\hat{m}_{2,0},)$ and its covariance matrix (Q_0 equals the identity matrix) and then follows a prediction-correction iteration. In the prediction step, the targeted parameters $m^f_{2,k}$ are predicted to be the same as the last estimates. In the correction step, the predicted closure $h(m^f_{2,k})$ using the predicted parameters $m^f_{2,k}$ is compared to the observation g_k to generate new estimates $m^a_{2,k}$. The iterative process is stopped when the average distances of the patient-specific model and the image based estimation between two consecutive iterations are less than 0.01mm or the maximum number of iteration is reached. The whole set of patient-specific parameters is obtained after the second step.

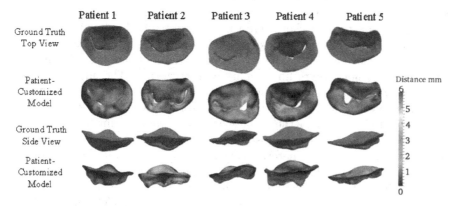

Fig. 2. The comparison between the patient-specific model and the image based estimation (treated as ground truth)

3   Experiments and Results

The automatic personalization algorithm is evaluated on the TEE images of five patients. First, the mitral valve apparatus and its motion is estimated from the TEE images using the machine learning method described in 2.2. The mitral leaflets are represented by tetrahedron finite elements with 9408 elements and 3248 vertices. Second, the mitral valve apparatus at the end diastole is loaded into the biomechanical model and the motion of mitral annulus and papillary tips are used as the prescribed boundary conditions. Third, the two step personalization algorithm is applied by adjusting the chordae rest length and material parameters from a coarse to fine level. The initial value of the chordae rest length is determined by the point-to-point distance from the papillary tip and the insertion points at the end systole. The initial value of the Young's modulus at the anterior and posterior leaflet is set to be 6.233 MPa and 2.087 MPa respectively which is the value estimated experimentally and used in [2].

Table 1. The Euclidean distances between the mitral valve closure generated from the biomechanical model and TEE with different level of personalization

	Chordae I	Chordae II	Chordae III	Chordae IV	Final	Semi-Manual
Patient 1	1.49±0.83	1.46±0.84	1.46±0.84	1.46±0.84	1.45±0.84	1.47±0.89
Patient 2	2.98±1.86	2.89±1.88	2.89±1.88	2.47±1.46	2.47±1.46	2.25±1.27
Patient 3	1.87±1.19	1.87±1.18	1.86±1.17	1.70±1.07	1.66±1.08	1.91±1.18
Patient 4	1.80±1.20	1.79±1.21	1.79±1.21	1.69±1.14	1.55±1.09	1.74±1.34
Patient 5	2.09±1.36	2.05±1.35	2.04±1.35	2.04±1.35	2.04±1.35	2.27±1.40

The results of the automatic personalization at each level compared to the semi-manual patient-customization method in [9] are shown in Table 1. The automatic algorithm performs similarly if not better than the semi-manual method with an expert adjusting the chordae rest length and the EKF adjusting the material parameters. The average fitting error is $1.84 \pm 1.17mm$. It can be seen that most patients achieve a good match at the coarse level of chordae adjustment. Some patients do not require fine tuning for the chordae. The adjustment of the chordae rest length brings the leaflet to the matching surface from the morphological perspective and reduces the average distances to about 2mm which is comparable to the error of the image-based observation from the quantitative perspective. The first step adjustment provides a better starting point to estimate the patient-specific material parameters to reduce the distance even further. Fig. 2 shows the distances between the personalized model and the image based estimation as the ground truth in the form of a color-map from both top and side views. It can be seen that the patient-specific model simulates the mitral valve closure very closely to image based estimation. The matching is especially close in the mitral annulus region thanks to the use of the boundary conditions. The performance of the algorithm can be improved in certain regions by employing the cost function with related terms (eq.1).

Table 2. Patient-specific material parameters

	E_{AL_f}	$E_{AL_{f_\perp}}$	G_{AL}	E_{PLf}	$E_{PL_{f_\perp}}$	G_{PL}
P1	6.28	2.37	2.11	2.21	1.99	0.74
P2	6.23	2.35	2.09	2.09	1.89	0.70
P3	5.73	2.16	1.93	4.58	4.14	1.54
P4	3.60	1.36	1.21	2.34	2.11	0.78
P5	6.23	2.35	2.09	2.09	1.89	0.70

The estimated patient-specific material parameters are shown in table 2. The anterior leaflet shows stiffer properties compared to the posterior leaflet for all patients. The general material parameters are also the optimized estimation for two patients. Different initial value of Young's modulus are used here but reach the same estimate.

4   Discussion and Conclusion

An automatic personalization algorithm is presented to estimate the patient-specific chordae rest length and material parameters. The quantitative evaluation on five patients demonstrates that the algorithm works as an efficient tool for patient-specific biomechanical model calibration. The algorithm allows the biomechanical model to simulate mitral valve closure with an average Euclidean distance of $1.84 \pm 1.17mm$ compared to the image based observation without any user interaction. It provides possibilities for biomechanical model to be used in clinical applications to simulate mitral valve motion and surgical modification in a user-friendly setting. The closer match between the biomechanical model and image observation creates a solid foundation for predictive surgical simulations. In addition, the patient-specific material parameters estimated using the algorithm may be used as quantitative support to explain the pathological changes. The applications of the algorithm are not limited to one imaging modality or certain biomechanical models. Instead, it provides a framework with great flexibility which can be used for cardiac Computed Tomography (CT) and Magnetic Resonance (MR) images with a different biomechanical model. Further validation on animal data compared to the material parameters measured in-vivo will confirm the performance of the algorithm.

References

1. Votta, E., Le, T.B., Stevanella, M., Fusini, L., Caiani, E.G., Redaelli, A., Sotiropoulos, F.: Toward patient-specific simulations of cardiac valves: State-of-the-art and future directions. J. Biomech. 46(2), 217–228 (2013)
2. Mansi, T., Voigt, I., Georgescu, B., Zheng, X., Mengue, E.A., Hackl, M., Ionasec, R.I., Noack, T., Seeburger, J., Comaniciu, D.: An integrated framework for finite-element modeling of mitral valve biomechanics from medical images: application to MitralClip intervention planning. Med. Image Anal. 16(7), 1330–1346 (2012)
3. Schneider, R.J., Tenenholtz, N.A., Perrin, D.P., Marx, G.R., del Nido, P.J., Howe, R.D.: Patient-specific mitral leaflet segmentation from 4D ultrasound. In: Fichtinger, G., Martel, A., Peters, T. (eds.) MICCAI 2011, Part III. LNCS, vol. 6893, pp. 520–527. Springer, Heidelberg (2011)
4. Schneider, R., Perrin, D., Vasilyev, N., Marx, G., del Nido, P., Howe, R.: Mitral annulus segmentation from four-dimensional ultrasound using a valve state predictor and constrained optical flow. Medical Image Analysis (2011)
5. Ionasec, R.I., Voigt, I., Georgescu, B., Wang, Y., Houle, H., Vega-Higuera, F., Navab, N., Comaniciu, D.: Patient-specific modeling and quantification of the aortic and mitral valves from 4-D cardiac CT and TEE. IEEE Trans. Med. Imaging 29(9), 1636–1651 (2010)
6. Voigt, I., Mansi, T., Ionasec, R.I., Mengue, E.A., Houle, H., Georgescu, B., Hornegger, J., Comaniciu, D.: Robust physically-constrained modeling of the mitral valve and subvalvular apparatus. In: Fichtinger, G., Martel, A., Peters, T. (eds.) MICCAI 2011, Part III. LNCS, vol. 6893, pp. 504–511. Springer, Heidelberg (2011)
7. Wang, Q., Sun, W.: Finite element modeling of mitral valve dynamic deformation using patient-specific multi-slices computed tomography scans. Ann. Biomed. Eng. (July 18, 2012)

8. Shi, P., Liu, H.: Stochastic finite element framework for simultaneous estimation of cardiac kinematic functions and material parameters. Med. Image Anal. 17(4), 445–464 (2003)
9. Kanik, J., Mansi, T., Voigt, I., Sharma, P., Ionasec, R., Comaniciu, D., Duncan, J.: Estimation of Patient-specific Material Properties of the Mitral Valve Using 4D Transesophageal Echocardiography. In: Proceeding of IEEE International Symposium on Biomedical Imaging (ISBI) (2013)

Fast Catheter Tracking in Echocardiographic Sequences for Cardiac Catheterization Interventions

Xianliang Wu[1], R. James Housden[2], Niharika Varma[2], YingLiang Ma[2], Kawal S. Rhode[2], and Daniel Rueckert[1]

[1] Department of Computing, Imperial College London, SW7 2AZ, UK
[2] Division of Imaging Sciences and Biomedical Engineering, King's College London, SE1 7EH, UK

Abstract. For most cardiac catheterization interventions, X-ray imaging is currently used as a standard imaging technique. However, lack of 3D soft tissue information and harmful radiation mean that X-ray imaging is not an ideal modality. In contrast, 3D echocardiography can overcome these disadvantages. In this paper, we propose a fast catheter tracking strategy for 3D ultrasound sequences. The main advantage of our strategy is low use of X-ray imaging, which significantly decreases the radiation exposure. In addition, 3D soft tissue imaging can be introduced by using ultrasound. To enable the tracking procedure, initialization is carried out on the first ultrasound frame. Given the location of the catheter in the previous frame, which is in the form of a set of ordered landmarks, 3D Speeded-Up Robust Feature (SURF) responses are calculated for candidate voxels in the surrounding region of each landmark on the next frame. One candidate is selected among all voxels for each landmark based on Fast Primal-Dual optimization (Fast-PD). As a result, a new set of ordered landmarks is extracted, corresponding to the potential location of the catheter on the next frame. In order to adapt the tracking to the changing length of the catheter in the view, landmarks which may not be located on the catheter are ruled out. Then a catheter growing strategy is performed to extend the tracked part of the catheter to the untracked part. Based on 10 ultrasound phantom sequences and two clinical sequences, comprising more than 1300 frames, our experimental results show that the tracking system can track catheters with an error of less than 2.5mm and a speed of more than 3 fps.

Keywords: Catheter Tracking, Ultrasound, Cardiac Catheterization.

1 Introduction

X-ray fluoroscopic imaging is used as a standard modality in cardiac catheter ablation procedures. However, it lacks 3D soft tissue information and uses harmful ionizing radiation. Alternatively, electroanatomical mapping systems are able to locate and guide catheters relative to the anatomy, but require additional hardware to track the catheters. In contrast, echocardiographic (ultrasound, US)

O. Camara et al. (Eds.): STACOM 2013, LNCS 8330, pp. 171–179, 2014.
© Springer-Verlag Berlin Heidelberg 2014

imaging has none of these problems and has been used to guide other types of interventions such as needle biopsies. The main disadvantages of US are its narrow field of view and acoustic artifacts when visualizing interventional devices. These drawbacks make detecting and tracking these devices challenging. Quantitative localisation can be used to augment visualisation or to monitor the precise treatment sites relative to the soft tissue information. Some existing work has focused on general surgical tool tracking in US by introducing external markers. In [1,2], passive markers attached to a surgical instrument were used to estimate the position and orientation of the instrument based on the US image using simple image processing. In [3], small US sensors were mounted on surgical tools to receive and transform acoustic energy to electrical signals, which were then analyzed to reconstruct the 3D coordinates of the tool. However, external markers are difficult to attach to micro surgical tools such as thin catheters. Micro surgical tool tracking in US has mainly focused on biopsy needle tracking. A efficient and robust needle detection and tracking algorithm based on level sets and partial differential equations was proposed in [4]. A more general micro tool localization in US based on model fitting using random sample consensus (RANSAC) and local optimization refinement was proposed in [5] and [6]. Recently another improved needle detection algorithm was proposed in [7]. Kalman filtering was used to combine the results of RANSAC-based and speckle tracking-based needle tip localization.

Currently, most research is focused on tracking rigid surgical tools rather than non-rigid tools such as long, flexible and thin catheters. To the best of our knowledge, only [8] managed to extract catheters in US volumes. The key idea was to use a registration between X-ray and US imaging. The tracking results in X-ray images were then employed to extract catheters from a reduced search space in US. This work had limited experimental validation and required long X-ray exposures.

In this paper, we present a fast catheter tracking system based on only ultrasound imaging for cardiac catheterization interventions. The main contributions of this work are as follows: (1) To the best of our knowledge, this is the first work towards real-time catheter tracking using only ultrasound sequences for cardiac catheterization; (2) Compared with combined X-ray and US guided catheter tracking [8], we perform experimental validation in more than 800 phantom and more than 500 clinical US images. Our results show that with a small trade-off in accuracy we can avoid long-term X-ray exposure and achieve higher tracking speed.

2   Methodology

Given an initialization comprising a set of ordered landmarks defining the catheter on the first frame (obtained either manually or automatically[8]) the voxels in each landmark's neighborhood in the next frame are allocated with credits, which are calculated based on 3D Speed-Up Robust Features (SURF)[9]. Fast Primal-Dual optimization (Fast-PD) [10] is then used to select one voxel for each landmark. The selected and ordered voxels represent the potential catheter in the next frame. Non-credible voxels, which may not be located on the catheter, are removed. Finally, a catheter-growing algorithm is performed to extend the tracking to untracked parts of the catheter.

2.1 Search Neighborhood and Measurement Definition

The tracked catheter in US is represented as a group of ordered landmarks. The distance between each pair of adjacent landmarks is 3–5 voxels (approximately 1–2 mm). Given the representation of the catheter in the previous frame, the search for the catheter in the next frame is constrained to a 2D search region around each landmark, each with its own local coordinate system. Within this region, each candidate voxel is allocated with a credit, calculated based on the SURF feature response in 3D.

Search Neighborhood Definition Based on Local Coordinate System. Each landmark $\mathbf{p}_i, i = 1, \ldots, n$ ($\mathbf{p}_i$ denotes the global coordinate in the US volume) defines a local coordinate system $O_{l,i}X_{l,i}Y_{l,i}Z_{l,i}$. The direction of axis $Z_{l,i}$ is aligned tangentially to the catheter, towards the next landmark. The direction of $X_{l,1}$ and $Y_{l,1}$, for the first landmark, can be determined arbitrarily. However, given the coordinate system of a previous landmark i, $X_{l,i+1}$ and $Y_{l,i+1}$ should be determined according to the following rule, shown in Fig. 1.

Given the local coordinate system $O_{l,i}X_{l,i}Y_{l,i}Z_{l,i}$ (colored in blue in Fig. 1(a)) of the previous landmark and the direction of $Z_{l,i+1}$ (colored in green) for the next landmark, we first rotate $O_{l,i}X_{l,i}Y_{l,i}Z_{l,i}$ around axis $Y_{l,i}$ to align the Z axis to the projection of axis $Z_{l,i+1}$ onto the plane $O_{l,i}X_{l,i}Z_{l,i}$. This defines a new coordinate system $O_{temp}X_{temp}Y_{temp}Z_{temp}$ (colored in red). This is then rotated around $O_{temp}X_{temp}$ to align axis Z_{temp} with axis $Z_{l,i+1}$. Through these two rotations, the local coordinate system $O_{l,i+1}X_{l,i+1}Y_{l,i+1}Z_{l,i+1}$ is determined. The purpose of this transformation is to ensure that two voxels which have the same local coordinates but in two adjacent coordinate systems have a small spatial distance between them. This is shown in red in Fig. 1(b) and (c).

After the definition of the local coordinate system for each landmark, the search region for landmark $\mathbf{p}_i, i = 1, \ldots, n$ can be denoted by:

$$S_i = \{|x_{l,i}| < R, |y_{l,i}| < R, z_{l,i} = 0\} \tag{1}$$

where $(x_{l,i}, y_{l,i}, z_{l,i})$ are the local coordinates and R is a search range. A more efficient discrete search is realized by constraining the search range to the plane $z_{l,i} = 0$ for each landmark.

Measurement Definition. For each voxel $c_{i,j}, j = 1, \ldots, m$ in the search region for a landmark $\mathbf{p}_i$, a credit $cred_{i,j}$ is allocated. The credit indicates the likelihood of the catheter passing through this voxel. Hessian matrix based methods are preferred to calculate the credit. We use 3D SURF because it is fast, based on integral images and robust to inaccurate computation of the Hessian matrix compared with the Frangi vesselness filter[11].

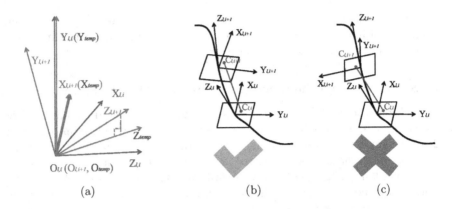

Fig. 1. Local coordinate definition. (a) Transformation between two local coordinate systems for adjacent landmarks; (b) Correct determination of local coordinates for the next landmark; (c) Incorrect determination of local coordinates for the next landmark.

2.2   Fast-PD Optimization

Fast-PD [10] is a fast discrete multi-label approach for solving a Markov Random Field (MRF) optimization based on a weighted graph. In our application, the graph is defined as $G =< V, E >$, where V consists of all of the n landmarks of the catheter and E consists of edges connecting any two adjacent landmarks. For each landmark $\mathbf{p}_i$, a search region S_i and the corresponding local coordinate system $O_{l,i}X_{l,i}Y_{l,i}Z_{l,i}$ are defined according to Section 2.1, with each region S_i containing m voxels. Each voxel $c_{i,j}$ has its own credit $cred_{i,j}$ and its own local 2D coordinate $(x_{l,i,j}, y_{l,i,j})$ (because $z_{l,i,j} = 0$, all voxels in the search region are constrained to a local plane) corresponding to the ith landmark. Each coordinate in any local coordinate system within the search region is considered as one label. Voxels with the same 2D local coordinate, even though they belong to different coordinate systems corresponding to different landmarks, are assigned the same label. Thus all of the voxels in a search region define a label space L of size m.

The energy function in our application is defined as:

$$\arg \min_{\mathbf{x}} f(\mathbf{x}) = \sum_{i=1}^{n}(u_i(x_i)) + \gamma \sum_{i=1}^{n-1}(dis(x_i, x_{i+1})) \qquad (2)$$

where $\mathbf{x} = \{x_1, x_2, \cdots, x_n\}$ is the selected voxel among all m candidates in the search region for each landmark, also known as the label assignment. The unary term $u_i(x_i)$ should be a decreasing function of its credit $cred_{i,j}$ (we used $u_i(x_i) = (1 - cred_{i,j})$). $dis(a, b)$ is the 2D Euclidean distance between the local 2D coordinates of the voxels corresponding to labels a and b.

Given $a, b \in L$, the unary term indicates the likelihood of the catheter passing through. The piecewise term is the planar distance between two locations in the search area corresponding to two labels, which functions as a smoothing term. All edge weights are set to one. γ is the balance parameter set manually.

2.3   Non-credible Landmark Removal and Catheter Growing

Fast-PD tracking can track catheters with motion along their normal direction, as this keeps the length of catheter in view unchanged. However, in practice the length of the catheter gradually changes. This gradual change causes the tracking error accumulate and eventually results in a tracking failure. To address this problem, a non-credible landmark removal algorithm, as well as a consequent catheter growing algorithm are included in the tracking system. Non-credible landmarks, which correspond to a low likelihood of a catheter, are ruled out to adapt the tracking to a decreasing length. If more of the catheter comes into view, the growing algorithm can extend to undetected parts.

Starting from the two ends of the already tracked part of the catheter, each landmark is checked for its credibility until a credible landmark is found. All non-credible landmarks are removed from the representation by a threshold τ_{cred} to each landmark's credit $cred_{i,lb}$, where lb is the label selected for landmark i through Fast-PD. After removal, an orientation $\mathbf{v}_i$, along which the growing should continue, is calculated for each of the two end landmarks i based on edge points extracted in the neighborhood. Because each edge point has its own gradient direction vector $\mathbf{e}_{i,j}, j = 1, \ldots, k$ (for k edge points), the orientation vector can be solved through the following equation:

$$\mathbf{A}_i \mathbf{v}_i = 0, \|\mathbf{v}_i\| = 1. \tag{3}$$

where $\mathbf{A}_i = \{\mathbf{e}_{i,1}^T, \mathbf{e}_{i,2}^T, \ldots, \mathbf{e}_{i,k}^T, \}^T$.

This equation is solved by SVD. After a coarse solution is obtained, non-supporting edge points, which have a smaller angle with $\mathbf{v}_i$, are removed (the threshold is $\pi/3$ in our application). Then the coarse solution is refined based on the remaining supporting points. The ratio of the number of supporting edge points over all edge points for each landmark is denoted by τ_i. If $\tau_i > 0.5$, the growing will continue, otherwise it is terminated.

If the orientation is accepted, then the step size of growing should be determined. Given two adjacent landmarks $\mathbf{p}_{i-1}$ and $\mathbf{p}_i$, a normalized vector $\mathbf{v}_{i-1,i}$ from $\mathbf{p}_{i-1}$ to $\mathbf{p}_i$ is calculated. The size of the growth step is determined by:

$$\triangle s_i = (M - N)\mathbf{v}_{i-1,i}^T \mathbf{v}_i + N. \tag{4}$$

where M and N are the maximum and minimum size of growth step set manually ($M = 5, N = 3$ voxels in our application). By setting the size of the steps through this equation, the growth will take smaller steps when a sharp turn occurs. After both the direction and the size of the step are determined, the next candidate position can be acquired by:

$$\mathbf{p}_{i+1}' = \mathbf{p}_i + \triangle s_i(\mathbf{v}_i). \tag{5}$$

The final position of the next landmark is determined by finding the voxel within a defined neighborhood of $\mathbf{p}_{i+1}'$ with the largest credit calculated through the method in Section 2.1. If all of the candidates' credits are below the threshold τ_{cred}, the growth will be terminated.

3 Experiments

A comparison was made between the proposed algorithm with [8] using 10 sequences of dynamic phantom data with ground truth catheter positions manually annotated by a single expert observer. A further two long clinical sequences, containing cardiac valves with inserted guidewires (approximately 2mm in thickness), were also used to validate its practical performance. The phantom data, acquired with a Philips X7-2t TEE probe, comprised more than 800 frames with 160×64×208 voxels. The two clinical sequences comprised more than 500 frames in total with 144×160×208 and 160×64×208 voxels respectively. Experiments were performed on an Ubuntu Linux system on a 3.40GHz, 8GB desktop.

The following performance metrics were defined to evaluate each algorithm's speed, accuracy and robustness: (1) **Average frame rate.** This evaluates speed, which directly determines performance for real-time applications; (2) **Average tracking error.** For each landmark i, a shortest distance d_i to the ground truth is calculated. Then a threshold ρ is used to select correctly tracked landmarks with $d_i \leq \rho$. The average of d_i among correctly tracked landmarks is defined as the tracking error; (3) **Incorrect tracking percentage (ITP).** ITP is defined as the number of incorrectly tracked landmarks over the total number of landmarks. ITP indicates the reliability of the tracking results. (4) **Failed tracking percentage (FTP).** For each landmark on the ground truth, the minimum distance to the tracked curve is calculated. The landmark is considered as successfully tracked if the distance is below ρ. FTP is defined as the number of failed landmarks over the total number of landmarks on the ground truth. It evaluates to what extent the whole catheter can be tracked. (5) **Ratio of failed tracked frames.** If the average credit for all landmarks is below $\tau_{cred}/3$, we treat it as a failure. The number of frames where a tracking failure is detected over the total number of frames can indicate the need for re-initialisation.

Parameters were selected based on prior information and experiments on a range of candidate values. Some of them, such as the scale of SURF, are dependent on the data and were selected based on prior knowledge for each dataset such as the size of the catheter. Others were selected based on experiments and can be generally extended to other data. Based on clinical sequence 1, Fig. 2 shows the performance in terms of accuracy, ITP and FTP, which are the most important measures, for a range of values of the two most important parameters: the threshold for non-credible landmarks, τ_{cred}, and the balance ratio for Fast-PD, γ. These results show that a range of $\tau_{cred} = 0.2 \sim 0.4$ and $\gamma = 0.5 \sim 2$ gives robust performance. We set $\tau_{cred} = 0.3$ and $\gamma = 0.1$.

Fig. 3 shows a comparison with [8], marked with (x), for different ρ (1–15) based on all the phantom data. In addition, the average speed of [8] is 1.69 fps while ours is 6.68 fps. For phantom data, the ratios of failures are all zero. The results show that our method can achieve a faster speed and a decrease in FTP. The trade-off is a small rise in tracking error and ITP, which is caused by inferior echo imaging quality compared with X-ray imaging. The most important advantage of our method is that X-ray exposure can be reduced significantly.

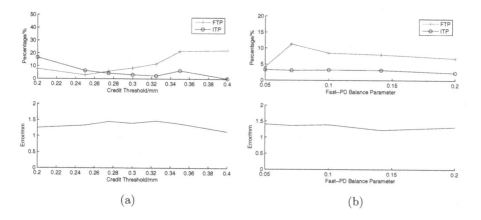

(a)   (b)

Fig. 2. Parameter selection based on clinical sequence 1:(a) feature credit threshold τ_{cred}; (b) Fast-PD balance γ

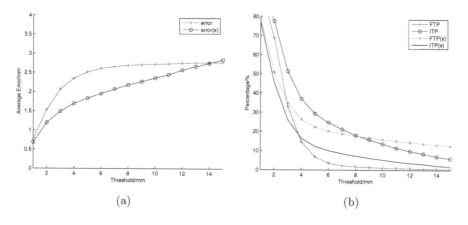

(a)   (b)

Fig. 3. Evaluation with different ρ based on phantom data

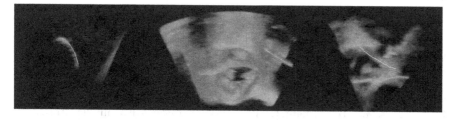

Fig. 4. An example of one tracked result. From left to right are phantom data, clinical sequence 01 and 02. The tracking results are marked with yellow curves.

Given $\rho = 5mm$, our method on clinical sequences 01/02 can achieve a speed of 3.19/6.19 fps and an error of 1.39/2.36 mm, with FTP of 8.47/10.96%, ITP of 3.21/10.73% and a failure ratio of 0.00/6.13%, respectively. Fig. 4 shows an example of the tracked results on both phantom and clinical data.

4   Conclusion

We have proposed a fast catheter tracking strategy based on only ultrasound imaging for cardiac catheterization. The experimental results show that it can track catheter motions at more than 3fps, with an error of less than 2.5mm. Less than 30% of the tracked results are incorrect and less than 11% of the ground truth is not tracked. Compared with [8], our method is faster and can track a larger percentage of the catheter with a small trade-off in accuracy and incorrectly tracked percentage. The low failure rate within 7% indicates that re-initialisation by X-ray or manual intervention will be rare. Future work will introduce catheter tip localization and temporal consistency into this system.

References

1. Stoll, J., Dupont, P.: Passive markers for ultrasound tracking of surgical instruments. In: Duncan, J.S., Gerig, G. (eds.) MICCAI 2005. LNCS, vol. 3750, pp. 41–48. Springer, Heidelberg (2005)
2. Novotny, P.M., Stoll, J.A., Vasilyev, N.V., del Nido, P.J., Dupont, P.E., Howe, R.D.: GPU based real-time instrument tracking with three dimensional ultrasound. In: Larsen, R., Nielsen, M., Sporring, J. (eds.) MICCAI 2006. LNCS, vol. 4190, pp. 58–65. Springer, Heidelberg (2006)
3. Mung, J., Vignon, F., Jain, A.: A non-disruptive technology for robust 3D tool tracking for ultrasound-guided interventions. In: Fichtinger, G., Martel, A., Peters, T. (eds.) MICCAI 2011, Part I. LNCS, vol. 6891, pp. 153–160. Springer, Heidelberg (2011)
4. Dong, B., Savitsky, E., Osher, S.: A novel method for enhanced needle localization using ultrasound-guidance. In: Bebis, G., et al. (eds.) ISVC 2009, Part I. LNCS, vol. 5875, pp. 914–923. Springer, Heidelberg (2009)
5. Uhercik, M., Kybic, J., Liebgott, H., Cachard, C.: Model fitting using RANSAC for surgical tool localization in 3-D ultrasound images. IEEE Transactions on Biomedical Engineering, 1907–1916 (2010)
6. Gaufillet, F., Liebgott, H., Uhercik, M., Cervenansky, F., Kybic, J., Cachard, C.: 3D ultrasound real-time monitoring of surgical tools. In: Ultrasonics Symposium (IUS), pp. 2360–2363 (2010)
7. Zhao, Y., Liebgott, H., Cachard, C.: Tracking micro tool in a dynamic 3D ultrasound situation using Kalman filter and RANSAC algorithm. In: ISBI, pp. 1076–1079 (2012)
8. Wu, X., Housden, J., Ma, Y., Rueckert, D., Rhode, K.S.: Real-time catheter extraction from 2D X-ray fluoroscopic and 3D echocardiographic images for cardiac interventions. In: Camara, O., Mansi, T., Pop, M., Rhode, K., Sermesant, M., Young, A. (eds.) STACOM 2012. LNCS, vol. 7746, pp. 198–206. Springer, Heidelberg (2013)

9. Bay, H., Ess, A., et al.: Speeded-up robust features (SURF). Computer Vision and Image Understanding, 346–359 (2008)
10. Komodakis, N., Tziritas, G., Paragios, N.: Fast, approximately optimal solutions for single and dynamic MRFs. In: CVPR, pp. 1–8 (2007)
11. Frangi, A.F., Niessen, W.J., Vincken, K.L., Viergever, M.A.: Multiscale vessel enhancement filtering. In: Wells, W.M., Colchester, A.C.F., Delp, S.L. (eds.) MICCAI 1998. LNCS, vol. 1496, pp. 130–137. Springer, Heidelberg (1998)

A Unified Statistical/Deterministic Deformable Model for LV Segmentation in Cardiac MRI

Sharath Gopal and Demetri Terzopoulos

Computer Science Department,
University of California, Los Angeles, USA

Abstract. We propose a novel deformable model with statistical and deterministic components for LV segmentation in cardiac magnetic resonance (MR) cine images. The statistical deformable component learns a global reference model of the LV using Principal Component Analysis (PCA) while the deterministic deformable component consists of a finite-element deformable surface superimposed on the reference model. The statistical model accounts for most of the global variations in shape found in the training set while the deterministic skin accounts for the local deformations consistent with the detailed image features. Intensity gradient-based image forces are applied to the model to segment and reconstruct LV shape. We validate our model on the MICCAI Grand Challenge dataset using leave-one-out training. Comparing the automated segmentation to the manual segmentation yields a Mean Perpendicular Distance (MPD) of 3.65 mm and a Dice coefficient of 0.86.

1 Introduction

The myocardial wall in the left ventricle (LV) is the main pumping structure of the heart and its function is important in the assessment of cardiovascular disease. By accurately segmenting the LV in cine MR images, cardiac contractile function can be quantified according to LV volumes and ejection fractions. Manual segmentation in MR images is a tedious process performed by clinicians, which is subject to inter- and intra-observer variability that can lead to inconsistent diagnosis. These issues have motivated researchers to develop automated methods that aspire to match the ability of expert clinicians to segment LV shape. The recent survey [1] provides an overview of different methods that have been applied to LV segmentation in MR images.

Deformable models have revolutionized model-based image analysis and their variational approach has been successfully applied to segmentation and tracking in medical images [2]. For example, the deformable model in [3] involves a regularization energy, which controls smoothness of the surface, and an image energy, which is generated from image features. The shape of the surface evolves under the influence of external forces to attain a minimum-energy configuration. Such models provide local control over the surface and include only weak shape priors that impose anatomical constraints yielding smooth, closed LV shapes in MR short axis images.

O. Camara et al. (Eds.): STACOM 2013, LNCS 8330, pp. 180–187, 2014.

The Deformable Superquadric Model (DSM) formulation in [4] includes a stronger shape prior in the form of a global superquadric shape. A finite-element locally deformable skin is superimposed on this global parameterized reference shape. The global and local degrees of freedom of the deformable model evolve under the influence of external forces, leading to the optimal fit of the reference shape and skin. As we show in the present paper, this formulation is appropriate for introducing statistically learned global reference shapes.

The statistical deformable model-based analysis of images was pioneered by the introduction of the Active Shape Model (ASM) [5] and the Active Appearance Model (AAM) [6], which use Principal Component Analysis (PCA) to learn appropriate shape/deformation priors from hand-segmented data. The AAM has been successfully applied to LV segmentation from cardiac MR images [7–10]. These methods provide a strong prior for shape and texture, such that the resulting shape is influenced by the variations present in the training set. PCA-based methods provide global control of shape and appearance, but the PCA priors can often be too restrictive and may not generalize well to variations not observed in the training set. Such uncommon variations can occur in pathological cases such as myocardial infarction and cardiomyopathy.

The above considerations have motivated our efforts to combine statistical and deterministic deformable models. In [11] we combined AAMs and DSMs in a multi-model, multi-stage approach. In the present paper, we further develop our approach by proposing a novel unified deformable model that replaces the superquadric reference shape with a statistically-learned PCA reference shape.

With similar motivations, a PCA shape prior is embedded in the internal energy formulation of a deformable mesh in [13]. At each iteration, the internal energy is dictated by the projection of the resulting shape on the modes of variation. The gradient-based data term is part of the external energy, and is responsible for pulling the mesh towards features of interest. Our model is more generic in the sense that any kind of external forces (gradient, inertial, or optical flow) can be applied to influence shape without any change to the model's formulation. The PCA parameters evolve simultaneously along with the pose and local displacement parameters, all under the influence of external forces. Such abstraction of external forces facilitates the design of image potential functions that influence the shape of the model. For example, we can design a potential based on optical flow to change the shape of the model according to the LV motion across phases in the images. Finally, the finite element skin provides good control over the smoothness of the surface, and by virtue of its ability to evolve independently, it is able to assume shapes that may not have been captured by the learned deformations of the PCA reference shape.

2   Model Formulation

Our formulation combines the strengths of deterministic and statistical deformable models. On the one hand, a PCA statistical component provides a strong learned shape prior that captures the global variations in the training

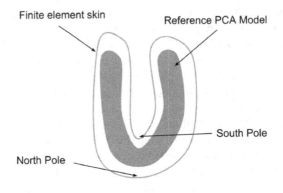

Fig. 1. Model Geometry

set. On the other hand, a deterministic deformable component affords local variations in shape that are dictated by the observed image features. The formulation of our model mainly involves embedding the PCA reference as illustrated in Fig. 1.

2.1 Geometry

The model is a closed surface that has $\mathbf{u} = (u, v)$ as its material coordinates. Principal Component Analysis (PCA) is applied to a set of aligned 3D LV shapes to obtain a discrete reference shape $\mathbf{s}(\mathbf{u})$ as

$$\mathbf{s}(\mathbf{u}) = \bar{\mathbf{s}}(\mathbf{u}) + \mathbf{P}_s(\mathbf{u})\mathbf{q}_s(\mathbf{u})^T, \tag{1}$$

where $\bar{\mathbf{s}}$ is the mean shape, the columns of matrix $\mathbf{P}_s$ are the modes of variation, and $\mathbf{q}_s$ are the shape or the PCA parameters. The translational offsets across the training shapes are removed by translating the respective centroids to the origin, and the rotational offsets are removed using Ordinary Procrustes Analysis. The Jacobian of the PCA reference shape $\mathbf{s}$ is given by

$$\mathbf{J}(\mathbf{u}) = \frac{\partial \mathbf{s}(\mathbf{u})}{\partial \mathbf{q}_s(\mathbf{u})} = \mathbf{P}_s(\mathbf{u}), \tag{2}$$

thus characterizing how the shape changes when the parameters $\mathbf{q}_s$ change. The Jacobian is key to the interaction of external forces with the model dynamics described later.

A finite element deformable skin is superimposed on the reference shape (Fig. 2) to account for local deformations. The local displacements $\mathbf{d}(\mathbf{u})$ are expressed as a linear combination of finite element basis functions $\mathbf{b}_i(\mathbf{u})$ as follows:

$$\mathbf{d}(\mathbf{u}) = \sum_i \mathrm{diag}(\mathbf{b}_i(\mathbf{u}))\mathbf{q}_i = \mathbf{S}\mathbf{q}_d, \tag{3}$$

where $\mathbf{q}_d = (..., \mathbf{q}_i, ...)^T$ is a set of local displacements $\mathbf{q}_i$ at each mesh node i and $\mathbf{S}$ holds the basis functions. In addition to the PCA parameters $\mathbf{q}_s$ and the local displacement parameters $\mathbf{q}_d$, the unified model also has global translation and rotation parameters $\mathbf{q}_c$ and $\mathbf{q}_\theta$. All the degrees of freedom (DOF) for the model are collected in a single vector

$$\mathbf{q} = (\mathbf{q}_c^T, \mathbf{q}_\theta^T, \mathbf{q}_s^T, \mathbf{q}_d^T)^T. \tag{4}$$

2.2 Dynamics

Given a new set of MR image slices for a patient, the vector $\mathbf{q}$ yielding a model that best fits the images must be computed. Applying Lagrangian dynamics, the model is made dynamic in $\mathbf{q}$, thus characterizing the evolution of $\mathbf{q}$ under the influence of external forces. The equations of motion are given as

$$\mathbf{C}\dot{\mathbf{q}} + \mathbf{K}\mathbf{q} = \mathbf{f}_q, \tag{5}$$

where $\dot{\mathbf{q}}$ is the time derivative of the DOF, $\mathbf{C}\dot{\mathbf{q}}$ are damping forces, $\mathbf{K}\mathbf{q}$ are elastic forces and $\mathbf{f}_q$ are external forces applied to the model. The stiffness matrix $\mathbf{K}$ determines the material/elastic properties of the finite element skin.

We impose a spline deformation energy on the local displacements $\mathbf{q}_d$ as

$$E(\mathbf{d}) = \int w_1(\mathbf{u}) \left(\left(\frac{\partial \mathbf{d}}{\partial u}\right)^2 + \left(\frac{\partial \mathbf{d}}{\partial v}\right)^2 \right) + w_0(\mathbf{u})\mathbf{d}^2 \, du \, dv, \tag{6}$$

where $w_0(\mathbf{u})$ controls the magnitude of the local deformation and $w_1(\mathbf{u})$ controls its variation across adjacent nodes on the skin.

The equations of motion in (5) are integrated through time using an explicit Euler method. The degrees of freedom in the vector $\mathbf{q}$ are updated from time t to time $t + \Delta t$ as follows:

$$\mathbf{q}^{(t+\Delta t)} = \mathbf{q}^{(t)} + \Delta t \left(\mathbf{C}^{(t)}\right)^{-1} \left(\mathbf{f}_q^{(t)} - \mathbf{K}\mathbf{q}^{(t)}\right). \tag{7}$$

Such a system will come to rest when the internal (damping and elastic) and external (image) forces equilibrate. Additional background details about the formulation and implementation are provided in [4].

2.3 Image Forces

The model, initialized with a mean PCA reference shape, is placed in the 3D volume formed by the MR slice stack, and deformed under the influence of image forces. The image forces are designed to attract the surface of the model towards the respective myocardial boundaries (endo and epi). The first step in designing forces is to define a gradient-based potential function on an image I as

$$P = \|\nabla(G_\sigma * I)\|, \tag{8}$$

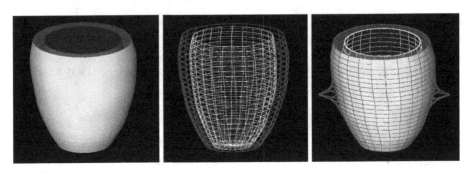

Fig. 2. PCA reference (left). Finite Element skin (center). Skin pulled away from the reference (right).

where the Gaussian smoothing width σ determines the range of influence of the forces. Multiple smoothing widths (Fig. 3) are used to attain equilibrium faster. Such a potential function presents image forces that attract the surface of the model towards image intensity edges. The force distribution is the gradient of the potential function:

$$\mathbf{f} = \beta \nabla P, \tag{9}$$

where β controls the scale of the force. The values for β, the stiffness parameters $w_1(\mathbf{u})$ and $w_0(\mathbf{u})$, and the time step Δt are carefully selected to maintain stability. In our implementation, we have used constant values $w_1 = 4 \times 10^{-3}$ and $w_0 = 2 \times 10^{-6}$, $\beta = 30$, and $\Delta t = 1$. The image smoothing and normalization methods affect the choice of these values.

We apply two different kinds of forces to the inner and outer walls of our model to differentiate between the endocardial wall (LV blood pool-myocardium interface) and the epicardial wall (myocardium/right ventricle (RV) and myocardium/outer organs interfaces). Since the blood pool and the pericardial fat appear bright and the myocardium appears dark in cine MR images, we can make use of the information present in the direction of the image gradients. At the endocardial border, the image gradients are oriented towards the LV blood pool, whereas at the epicardial border, the image gradients are oriented away from the LV blood pool. Thus, the endocardial forces $\mathbf{f}_I$ and the epicardial forces $\mathbf{f}_o$ are given as

$$\mathbf{f}_I(\mathbf{u}) = \begin{cases} \mathbf{f}(\mathbf{u}), & \text{if } \nabla I \cdot \mathbf{x}(\mathbf{u}) < 0 \\ 0, & \text{otherwise,} \end{cases} \tag{10}$$

$$\mathbf{f}_o(\mathbf{u}) = \begin{cases} \mathbf{f}(\mathbf{u}), & \text{if } \nabla I \cdot \mathbf{x}(\mathbf{u}) > 0 \\ 0, & \text{otherwise,} \end{cases} \tag{11}$$

where the tests involve the projection of the image gradient on the position vectors $\mathbf{x}(\mathbf{u})$ of the points on the model surface whose centroid is at the origin.

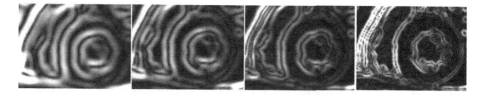

Fig. 3. Image potentials at multiple smoothing widths (4, 3, 2, 0 mm)

Table 1. Mean Perpendicular Distance (mm) and Dice coefficient (45 cases)

MPD	ED-Epi	ED-Endo	Dice	ED-Epi	ED-Endo
MEAN	3.6	3.7	MEAN	0.88	0.84
STD	0.52	0.62	STD	0.02	0.04
MAX	4.89	4.68	MAX	0.93	0.91
MIN	2.11	2.16	MIN	0.81	0.75

3 Results

We validated the segmentation ability of our model using leave-one-out train-ing on end-diastolic (ED) images of the 45 MICCAI Grand Challenge datasets. The leave-one-out validation was fully automated and the mean reference model was initialized in the volume such that the centroid of the model coincided with the center of the mid-slice. Initially, the model was subject only to translational forces designed using optical flow potentials across phases. Such forces approx-imately localize the myocardium and help in moving the initial mean reference closer to the actual solution. Subsequently, all the parameters (rigid and non-rigid) were stepped forward in time. The PCA parameters q_s are restricted within +2 and −2 standard deviations (which can be obtained from the corre-sponding eigenvalues) from the mean in order to prevent unlikely shapes. The deformation of the skin is controlled by the w_1 and w_0 constants. The model is stepped forward across multiple Gaussian smoothing widths (4, 3, 2, 0 mm), finally converging at the myocardial boundaries.

Due to the ambiguous gradient information at the myocardium interface with lungs and other organs, the epicardial boundary is harder to localize. By virtue of the model having the MICCAI Grand Challenge trained PCA reference shape, the papillary muscles were included in the blood pool (Fig. 4). We used the Mean Perpendicular Distance (MPD) and Dice coefficients (Table 1) to compare the positioning errors of automated contours with respect to the expert-delineated contours. The average Dice coefficient and the average MPD for the ED seg-mentation are 0.86 and 3.65 mm respectively, and these are close to the results presented in [14], [15], and [16]. The automated contours for the mid-slices are more accurate than those for the slices towards the apex due to partial volume effects.

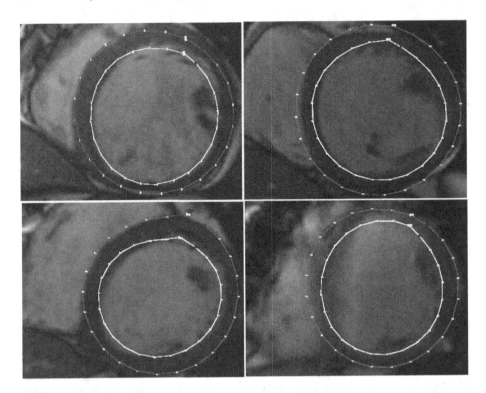

Fig. 4. Examples of automated contour segmentation for four cases

4   Conclusion

We have proposed a novel deformable model for LV segmentation in cardiac MR images, which combines a strong statistical prior learned from manually-segmented training data through PCA with a finite element deformable skin. Our model is unique in the sense that the PCA reference shape is embedded in its physical formulation and it evolves under the influence of external image forces. Leave-one-out validation on the 45 MICCAI Grand Challenge datasets yields good results. Our work brings us a step closer to the automation of all phases of LV segmentation in cardiac cine MRI.

References

1. Petitjean, C., Dacher, J.: A review of segmentation methods in short axis cardiac MR images. Medical Image Analysis 15, 169–184 (2011)
2. McInerney, T., Terzopoulos, D.: Deformable models in medical image analysis: A survey. Medical Image Analysis 1(2), 91–108 (1996)
3. McInerney, T., Terzopoulos, D.: A dynamic finite element surface model for segmentation and tracking in multidimensional medical images with application to 4-D image analysis. Comp. Med. Imag. Grap. 19(1), 69–83 (1995)

4. Terzopoulos, D., Metaxas, D.: Dynamic 3D models with local and global deformations: Deformable superquadrics. IEEE Transactions on PAMI 13(7), 703–714 (1991)
5. Cootes, T.F., Taylor, C.J., Cooper, D.H., Graham, J.: Active shape models: Their training and application. Comp. Vis. Imag. Understanding 61, 38–59 (1995)
6. Cootes, T.F., Edwards, G.J., Taylor, C.J.: Active appearance models. In: Burkhardt, H., Neumann, B. (eds.) ECCV 1998. LNCS, vol. 1407, pp. 484–498. Springer, Heidelberg (1998)
7. Mitchell, S.C., Bosch, J.G., Lelieveldt, B.P.F., van der Geest, R.J., Reiber, J.H.C., Sonka, M.: 3D Active appearance models: Segmentation of cardiac MR and ultrasound images. IEEE Transactions on Medical Imaging 21(9), 1167–1178 (2002)
8. Mitchell, S.C., Lelieveldt, B.P.F., van der Geest, R.J., Bosch, H.G., Reiber, J.H.C., Sonka, M.: Multistage hybrid Active Appearance Model matching: Segmentation of left and right ventricles in cardiac MR images. IEEE Transactions on Medical Imaging 20(5), 415–423 (2001)
9. Zhang, H., Wahle, A., Johnson, R.K., Scholz, T.D., Sonka, M.: 4-D cardiac MR image analysis: Left and right ventricular morphology and function. IEEE Transactions on Medical Imaging 29(2), 350–364 (2010)
10. Stegmann, M.B., Perdersen, D.: Bi-temporal 3D active appearance models with applications to unsupervised ejection fraction estimation. Proc. of SPIE Medical Imaging, 5747, 336–350 (2005)
11. Gopal, S., Otaki, Y., Arsanjani, R., Berman, D., Terzopoulos, D., Slomka, P.: Combining active appearance and deformable superquadric models for LV segmentation in cardiac MRI. Proc. SPIE Medical Imaging, 8669-15:1–8 (2013)
12. Radau, P., Lu, Y., Connelly, K., Paul, G., Dick, A.J., Wright, G.A.: Evaluation framework for algorithms segmenting short axis cardiac MRI. The MIDAS Journal - Cardiac MR Left Ventricle Segmentation Challenge (2009), http://hdl.handle.net/10380/3070
13. Kaus, M., Von Berg, J., Weese, J., Niessen, W., Pekar, V.: Automated segmentation of the left ventricle in cardiac MRI. Medical Image Analysis 8(3), 245–254 (2004)
14. Jolly, M.: Fully automatic left ventricle segmentation in cardiac cine MR images using registration and minimum surfaces. The MIDAS Journal—Cardiac MR Left Ventricle Segmentation Challenge (2009)
15. Huang, S., Liu, J., Lee, L., Venkatesh, S., Teo, L., Au, C., Nowinski, W.: Segmentation of the left ventricle from cine MR images using a comprehensive approach. The MIDAS Journal—Cardiac MR Left Ventricle Segmentation Challenge (2009)
16. Lu, Y., Radau, P., Connelly, K., Dick, A., Wright, G.: Automatic image-driven segmentation of left ventricle in cardiac cine MRI. The MIDAS Journal—Cardiac MR Left Ventricle Segmentation Challenge (2009)

Multi-modal Pipeline for Comprehensive Validation of Mitral Valve Geometry and Functional Computational Models

Dominik Neumann[1,2], Sasa Grbic[1,3], Tommaso Mansi[1], Ingmar Voigt[1],
Jean-Pierre Rabbah[4], Andrew W. Siefert[4], Neelakantan Saikrishnan[4],
Ajit P. Yoganathan[4], David D. Yuh[5], and Razvan Ioan Ionasec[1]

[1] Imaging and Computer Vision, Siemens Corporate Technology, Princeton, NJ
[2] Pattern Recognition Lab, University of Erlangen-Nuremberg, Germany
[3] Computer Aided Medical Procedures, Technical University Munich, Germany
[4] The Wallace H. Coulter Department of Biomedical Engineering,
Georgia Institute of Technology and Emory University, Atlanta, GA
[5] Section of Cardiac Surgery, Department of Surgery,
Yale University School of Medicine, New Haven, CT
dominik.neumann@siemens.com

Abstract. Valvular heart disease affects a high number of patients, exhibiting significant mortality and morbidity rates. Mitral Valve (MV) Regurgitation, a disorder in which the MV does not close properly during systole, is among its most common forms. Traditionally, it has been treated with MV replacement. However, recently there is an increased interest in MV repair procedures, providing better long-term survival, better preservation of heart function, lower risk of complications, and usually eliminating the need for long-term use of blood thinners (anticoagulants). These procedures are complex and require an experienced surgeon and elaborate pre-operative planning. Hence, there is a need for efficient tools for training and planning of MV repair interventions. Computational models of valve function have been developed for these purposes. Nevertheless, state-of-the-art models remain approximations of real anatomy with considerable simplifications, since current modalities are limited by image quality. Hence, there is an important need to validate such low-fidelity models against comprehensive ex-vivo data to assess their clinical applicability. As a first step towards this aim, we propose an integrated pipeline for the validation of MV geometry and function models estimated in ex-vivo TEE data with respect to ex-vivo microCT data. We utilize a controlled experimental setup for ex-vivo imaging and employ robust machine learning and optimization techniques to extract reproducible geometrical models from both modalities. Using one exemplary case, we demonstrate the validity of our framework.

1 Introduction

The mitral valve (MV) separates the left atrium from the left ventricle and prevents the blood flow back to the left atrium during systole. Incorrect MV closure

O. Camara et al. (Eds.): STACOM 2013, LNCS 8330, pp. 188–195, 2014.

appears in many cardiac diseases and often requires MV replacement or repair surgery. In recent years, MV repair procedures, where the valve is surgically altered in order to restore its proper hemodynamic function, are substituting classical valve replacements [1]. It is the best option for nearly all patients with a regurgitant MV and for many with a narrowed (stenotic) MV [2]. However, the procedure is technically challenging and requires an experienced surgical team to achieve optimal results [1], since the deformation of complex valve anatomy during the intervention has to be predicted and associated with post-operative implications regarding valve anatomy and function. Having a framework to explore different surgical repair strategies for an individual patient and virtually compute their outcomes would be a desired tool in current clinical practice.

Driven by the growing prevalence of MV diseases, researchers are developing methods to assess MV anatomy from multiple imaging modalities and simulate its physiology using biomechanical models [3,4]. However, the clinical applicability is limited as they either do not enable patient-specific personalization of the geometric model or this process requires tedious manual interactions.

In recent years, methods have been proposed to delineate the MV using semi-manual or advanced automated algorithms [5]. Using these models, biomechanical computations can be performed based on a personalized patient-specific geometry as in [6]. However, their model relies on a simplified geometrical model, mainly due to the limitations of in-vivo TEE imaging. In particular, the MV leaflet clefts were not captured and the complex chordae anatomy was simplified with a parachute model. In order to apply such methods in clinical practice, the first step is to validate the prediction power of simplified models against ideal, high-fidelity models in a controlled ex-vivo environment.

We propose a novel validation framework for both geometric and biomechanical models extracted from non-invasive modalities. We developed a controlled experimental setup for MV ex-vivo imaging in order to acquire functional TEE data and high-resolution microCT images of the MV. Robust machine learning and optimization algorithms are utilized to produce accurate and reproducible models of the MV from both modalities. Based on the TEE model extracted during end-diastole (MV open), we utilize state of the art computational models to compute the geometric configuration of the MV during systole. We illustrate the capabilities of our framework using one excised sheep MV, showing promising results towards an integrated platform for comprehensive model validation.

2   Experimental Setup

In-vitro Simulator. A novel closed-loop left heart simulator (Fig. 1) was utilized to carry out controlled in-vitro MV experiments [7,8]. It allows for precise control of annular and subvalvular MV geometry at physiological left heart hemodynamics. The modular design was optimized to allow for micro-computed tomography (microCT) and echocardiography techniques. The rigid left atrium and ventricle consists of a thin-walled acrylic chamber with a cylindrical cross section, resulting in uniform scattering and X-ray absorption for the microCT

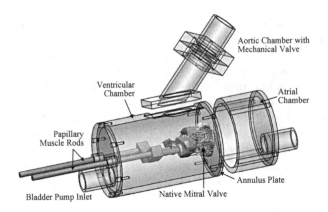

Fig. 1. Schematic representation of the in-vitro simulator with mounted mitral valve

and thus facilitating segmentation. The aortic chamber was designed at a 135° angle from the MV to preserve their physiologic spatial orientation.

Experimental Protocol. For experimentation, a fresh ovine heart was obtained and the MV excised preserving its annular and subvalvular anatomy. All leaflet chordae of the selected MV are inserting directly into the papillary muscles (PM). The MV was sutured to the simulator's annulus using a Ford interlocking stitch. During valve suturing, care was taken to place each suture just above the valve's natural hinge and not through the leaflet tissue.

After annular suturing, each PM was attached to the PM control rods. Each PM was carefully positioned and fine-tuned to establish the control MV geometry as previously described [9]. The simulator was filled with 0.9% saline solution and leaflet dynamics and coaptation geometry were studied at room temperature under physiologic hemodynamic conditions (120 mmHg peak left ventricular pressure, 5.0 L/min average cardiac output at a heart rate of 70 bpm) to ensure proper valvular function. 3D echocardiography at a good temporal resolution of 50 Hz was acquired using a Phillips iE33 system with an X7-2 pediatric probe.

MicroCT Protocol. The atrial chamber and aortic section were removed from the left heart simulator, and the left ventricle was fixed to the microCT gantry using a custom adaptor plate. The MV geometry did not get perturbed by these changes. The valve was scanned in air under ≈ 30 mmHg ventricular pressure using a viva CT 40 system (Scanco Medical AG). The geometry was acquired at 39 μm voxel size (≈ 600 slices) using scanning parameters optimized for low density soft tissues (55 keV energy, 109 uA intensity, 300 ms integration time).

3   Methods

3.1   Extraction of High-Fidelity Model from Ex-vivo MicroCT

We propose a novel automated segmentation procedure for microCT images of the MV (see Fig. 2), where the structure of the valve is decomposed into parts relevant to the subsequent computation of geometric measurements.

Fig. 2. Left: microCT scan of MV suspended in the simulator, right: extracted model

Locating MV Annulus and Papillary Muscles. An adaption of the approach proposed in [10] based on convex programming is exploited for the localization task. Efficient implementations as described in [11] allow for exhaustive parameter search within seconds. We utilize a sparse model of the MV, which can be generated manually or extracted automatically from corresponding TEE models. Instead of directly using the raw image, a smooth Euclidean distance transform of the MV is required. We approximate the transform by a distance map based on an anisotropically filtered image derived from the microCT scan using Danielsson's method [12]. The described workflow allows for the extraction of rough estimates of the centers of the papillary muscles, several points on the mitral annulus, and points in the posterior and anterior leaflet.

Extracting Full Model. Based on the estimated locations of the mitral annulus and papillary muscles, we generate seed points for the Random Walker algorithm, yielding a segmentation of the MV in the original image. The parameters were set heuristically. The resulting mask is then converted into a triangulated mesh model for the remaining model decomposition steps utilizing the Marching Cube algorithm.

Chordae Tracing. Given the MV mesh and points in each papillary muscle (PM) and on the annulus, we estimate the configuration of the chordae tendineae, which we use to determine the exact locations of the chordae insertion points, i.e. those locations, where a chord is connected to the leaflet. This information is necessary in order to split the MV mesh into PMs, chordae, annulus and leaflets. Chordae segmentation is not straightforward, since the chordae are structured in a tree-like fashion and a chord is not necessarily a straight line, it rather shows high curvature under certain conditions. Hence, we perform a novel path tracing approach based on geodesic distances on the mesh, where we determine all paths from the annulus to the PMs (start and end point are given by the sparse model estimation). It is a valid assumption that each of those paths contains a chord. Starting from the annulus, for each point on the descending path, we compute the mesh thickness at this location and retrieve the leaflet insertion points by analyzing the thickness profile throughout the path. A threshold of lower than 0.5 mm thickness on the descending path from the annulus was determined to be the location of an insertion point. Next, we cut the mesh at the insertion point and repeat the previous procedure until no more paths connecting the mitral annulus with one of the papillary muscles are found.

Fig. 3. Estimated MV model from ex-vivo TEE scan visualized in the TEE volume

3.2   TEE Mitral Valve Apparatus Parametrization and Estimation

In this study, we utilize the non-invasive anatomical point distribution model $\mathcal{S}$ of the MV and its subvalvular components from [5,13] estimated on 3D TEE. $\mathcal{S}$ comprises mitral annulus, and anterior and posterior leaflets and papillary tips. Nine anatomical landmarks (two trigones, two commissures, one posterior annulus mid-point, two leaflet tips, and two papillary tips) allow for a consistent, patient-specific derivation of $\mathcal{S}$, which is capable of capturing a broad spectrum of morphological variations. The parameters of $\mathcal{S}$ are incrementally estimated within the hierarchical, discriminative Marginal Space Learning framework using classifiers based on the Probabilistic Boosting Tree with Haar-like and Steerable features. Fig. 3 depicts the extracted model based on the TEE image.

3.3   Biomechanical Model of the Mitral Valve

MV closure is calculated from the TEE anatomy based on the model proposed in [6]. In brief, the dynamics system $M\ddot{u}+C\dot{u}+Ku = f_t+f_p+f_c$ is solved, where M is the diagonal mass matrix calculated from the mass density $\rho = 1040\,g/L$, C is the Rayleigh damping matrix with coefficients $1e4\,s^{-1}$ and $0.1\,s$ for the mass and stiffness matrix respectively, K is the stiffness matrix, f_t is the force created by the chords on the leaflets, f_p the pressure force, f_c the contact forces and u the displacement. In this study, we rely on transverse isotropic linear tissue elasticity implemented using a co-rotational finite elements method (FEM) to cope with large deformations. The choice of linear elasticity is motivated by recent findings suggesting a linear behavior in-vivo [14]. Poisson ratio is $\nu = 0.488$ for both leaflets, fiber Young's modulus is $E_{AL} = 6.23\,MPa$ and $E_{PL} = 2.09\,MPa$ for the anterior and posterior leaflets respectively, cross-fiber Young's modulus is $E_{AL} = 2.35\,MPa$ and $E_{PL} = 1.88\,MPa$ and shear modulus is $1.37\,MPa$. Mitral annulus and PMs are fixed. Chordae are modeled as in [6]: twenty-eight marginal chordae are evenly attached at the free-edges of the leaflets and four chordae are tethered at the base of the leaflets. In that way, we mimic a real-case scenario where the detailed configuration of the chordae is unknown. Chordae follow an exponential law [6]. Finally, self collisions are modeled, with collision stiffness of $100\,kPa$ and friction coefficient of 0.1. The model is implemented in the SOFA framework[1]. Spatial variables are discretized using linear tetrahedra while an Euler implicit time discretization is employed for robust computation.

---

[1] http://www.sofa-framework.org/

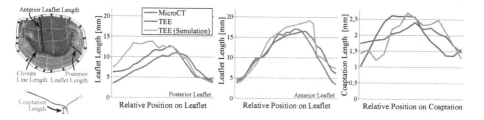

Fig. 4. Geometric comparison at systole from microCT, annotated and simulated TEE

4 Experimental Results

We utilize our framework on one ovine valve and compare the geometric config-uration between the model constructed from TEE and microCT during systole (MV closed). In addition, we compute the MV geometry at systole from an end-diastolic (MV open) TEE image and compare it to the ground-truth geometrical configuration obtained from the microCT image.

4.1 Geometric Comparison

Based on the geometric models extracted during systole (MV closed) from TEE and microCT, we measured clinically relevant parameters (coaptation length, coaptation area, closure line length, and anterior and posterior leaflet length, see Fig. 4, left panel) in order to quantitatively compare geometric differences between the two models (see Fig. 4, right panels). The maximum coaptation length is 2.41 mm vs 2.62 mm for microCT and TEE, respectively. Closure line lengths are measured as 23.08 mm vs 21.1 mm, and coaptation areas are 45.8 mm^2 vs 41.1 mm^2. Qualitative comparisons are shown in Fig. 5. These results indi-cate that the utilized simplified TEE model can accurately represent important biomarkers compared to an idealized model (extracted from microCT).

4.2 MV Closure Computation

Starting from the end-diastolic TEE MV model (last frame where the MV is seen open in the TEE image), we computed MV closure based on the model described in Sec. 3.3. In this experiment, we used a time step of 10 ms. A nominal pressure profile varying from 0 mmHg to 120 mmHg was applied [6]. Anterior leaflet chord rest length was set to 1.6× longer than the distance between free-edges and papillary muscles to cope with folded chords at end-diastole (MV open). That ratio was estimated on other microCT data. Finally, to capture the fast dynamics and correctly account for collisions and inertia, pressure increase duration was scaled to last 10 s. 1000 iterations were calculated.

Fig. 6 illustrates the calculated MV closure geometry from time 0 s to 7 s, before peak pressure was reached (afterwards the valve stayed closed). As one can see, our model could capture MV closure qualitatively well. Quantitatively,

Fig. 5. Qualitative comparison of MV geometry from TEE (cyan) and microCT (beige)

Fig. 6. Left: fiber orientations of computational TEE model, right: different time steps of computed MV closure geometry, initialized from end-diastolic (MV open) TEE image

we measured clinically-related parameters as described above from the predicted closure and compared them against the idealized model (Fig. 4). Maximum coaptation length is 2.71 mm for the computed model vs 2.41 mm for the idealized microCT model. Closure line lengths are measured as 23.3 mm vs 23.08 mm, and coaptation areas are 46.1 mm^2 vs 45.8 mm^2. These results confirm that simplified models from TEE can be utilized to build biomechanical models and compute accurate MV closure geometry in respect to relevant clinical parameters.

5   Conclusion

We proposed a novel complete pipeline for validating geometrical and functional models of the mitral valve utilizing a controlled ex-vivo setup capable of acquiring both high-resolution ex-vivo microCT scans in order to obtain ground-truth information, and standard low-resolution TEE images. This pipeline serves as a bridge between ex-vivo and non-invasive clinical modalities. We integrated robust algorithms in order to extract reproducible models from microCT and TEE images. Thus, quantitative comparison of clinically relevant measurements is possible. Measurements were computed from simplified models extracted from non-invasive modalities and compared to idealized high-fidelity models based on microCT. Finally, experiments on real data obtained from one ovine valve were conducted as a proof of concept, demonstrating the capabilities of our framework. Now that we have an integrated and comprehensive ex-vivo microCT / TEE setup, we are able to evaluate models in a consistent manner on a larger population and thus validate the prediction power of current in-vivo computational frameworks, which will be the next step of the current study.

References

1. Kilic, A., Shah, A., Conte, J., Baumgartner, W., Yuh, D.: Operative outcomes in mitral valve surgery: Combined effect of surgeon and hospital volume in a population-based analysis. J. Thorac. Cardiovasc. Surg. (2012)
2. Bolling, S., Li, S., O'Brien, S., Brennan, M., Prager, R., Gammie, J.: Predictors of mitral valve repair: Clinical and surgeon factors. Ann. of Thorac. Surg. 90, 1904–1912 (2013)
3. Wang, Q., Sun, W.: Finite element modeling of mitral valve dynamic deformation using patient-specific multi-slices computed tomography scans. Ann. Biomed. Eng. 41(1), 142–153 (2013)
4. Stevanella, M., Maffessanti, F., Conti, C., Votta, E., Arnoldi, A., Lombardi, M., Parodi, O., Caiani, E., Redaelli, A.: Mitral valve patient-specific finite element modeling from cardiac mri: Application to an annuloplasty procedure. Cardiovascular Engineering and Technology 2(2), 66–76 (2011)
5. Ionasec, R., Voigt, I., Georgescu, B., Wang, Y., Houle, H., Vega-Higuera, F., Nassir, N., Comaniciu, D.: Patient-specific modeling and quantification of the aortic and mitral valves from 4-D cardiac CT and tee. TMI 29(9), 1636–1651 (2010)
6. Mansi, T., Voigt, I., Georgescu, B., Zheng, X., Assoumou Mengue, E., Hackl, M., Ionasec, R.T., Seeburger, J., Comaniciu, D.: An integrated framework for fnite-element modeling of mitral valve biomechanics from medical images: Application to mitralclip intervention planning. Med. Image Anal. 16(7), 1330–1346 (2012)
7. Siefert, A.W., Rabbah, J.P.M., Koomalsingh, K.J., Touchton Jr., S.A., Saikrishnan, N., McGarvey, J.R., Gorman, R.C., Gorman III, J.H., Yoganathan, A.P.: In vitro mitral valve simulator mimics systolic valvular function of chronic ischemic mitral regurgitation ovine model. Ann. Thorac. Surg. 95, 825–830 (2013)
8. Rabbah, J.P., Saikrishnan, N., Yoganathan, A.P.: A novel left heart simulator for the multi-modality characterization of native mitral valve geometry and fluid mechanics. Ann. Biomed. Eng. 31, 305–315 (2012)
9. Jimenez, J.H., Soerensen, D.D., He, Z., He, S., Yoganathan, A.P.: Effects of a saddle shaped annulus on mitral valve function and chordal force distribution: an in vitro study. Ann. Biomed. Eng. 31(10), 1171–1181 (2003)
10. Chen, S.S., Donoho, D.L., Saunders, M.A.: Atomic decomposition by basis pursuit. SIAM 43(1), 129–159 (2001)
11. Breitenreicher, D., Schnoerr, C.: Model-based multiple rigid object detection and registration in unstructured range data. International Journal of Computer Vision 92(1), 1573–1405 (2011)
12. Danielsson, P.: Euclidean distance mapping. Computer Graphics and Image Processing 14(3), 227–248 (1980)
13. Voigt, I., Mansi, T., Ionasec, R.I., Mengue, E.A., Houle, H., Georgescu, B., Hornegger, J., Comaniciu, D.: Robust physically-constrained modeling of the mitral valve and subvalvular apparatus. In: Fichtinger, G., Martel, A., Peters, T. (eds.) MICCAI 2011, Part III. LNCS, vol. 6893, pp. 504–511. Springer, Heidelberg (2011)
14. Krishnamurthy, G., Itoh, A., Bothe, W., Swanson, J., Kuhl, E., Karlsson, M., Craig Miller, D., Ingels, N.: Stress–strain behavior of mitral valve leaflets in the beating ovine heart. J. Biomech. 42(12), 1909–1916 (2009)

Personalized Modeling of Cardiac Electrophysiology Using Shape-Based Prediction of Fiber Orientation

Karim Lekadir[1], Ali Pashaei[1], Corné Hoogendoorn[1], Marco Pereanez[1], Xènia Albà[1], and Alejandro F. Frangi[1,2]

Center for Computational Imaging & Simulation Technologies in Biomedicine
[1] Universitat Pompeu Fabra and CIBER-BBN, Barcelona, Spain
[2] University of Sheffield, Sheffield, United Kingdom

Abstract. Fibers play an important role in electrophysiological (EP) simulations as they determine the shape and directions of the electrical waves traveling throughout the myocardium. Due to the limited unavailability of *in vivo* images of the fiber structure, computational modeling of electrophysiology has been performed thus far mostly using the well-known rule-based Streeter model. The aim of this paper is to present an EP simulation study based on a statistics-based fiber model. With this approach, the missing subject-specific fiber model is predicted directly from the available shape information based on a predictive model constructed from a training sample of *ex vivo* DTI images. Experiments are carried out based on a database of canine datasets (including normal and abnormal cases), by considering the DTI-, the Streeter-, and the statistics-based fiber models. The results show that the shape-based predicted fiber models improve significantly the estimation accuracy of the electrical activation times and patterns, from average errors of about 10% to 1%.

1   Introduction

Simulation-based studies of many cardiac phenomena and pathologies, as well as the the *in silico* planning of related electrophysiological interventions, require the construction of accurate subject-specific models of fibers. This is because the fibers determine the shape and directions of the electrical waves traveling throughout the myocardium. Ideally, the subject-specific fibers would be extracted from *in vivo* DT-MRI, but the modality is known to be sensitive to the cardiac motion, which affects the diffusion measurements. As a result, alternative methods for the personalization of fiber orientations are still required in order to derive realistic EP models.

Thus far, the most popular method for generating fiber orientations remains the rule-based model by Streeter *et al.* ever since its introduction in 1969 [1]. It was originally described based on histological studies, before it was confirmed in more recent years using *ex vivo* DTI datasets [2]. Essentially, the Streeter model states that the fiber orientations rotate counterclockwise in the myocardium, while varying smoothly between the endo- and the epi-cardial walls. Currently, the Streeter model is by far the fiber model of choice in EP simulation as illustrated by these recent works (*e.g.*, [3], [4], [5], [6], [7], [8], [9], [10]). This model, however, is rather simplistic and

O. Camara et al. (Eds.): STACOM 2013, LNCS 8330, pp. 196–203, 2014.
© Springer-Verlag Berlin Heidelberg 2014

idealistic. While it approximates the overall patterns of the fiber structure, it might not conform well to the subject-specific local distribution of the myocytes. This can potentially lead to suboptimal simulations of the electrical activation.

As a result, some researchers have recently proposed alternative statistics-based techniques for the modeling of fibers, in order to encode the fiber variability across individuals and groups of individuals [11], [12]. In particular, the work in [12] suggests to predict the fiber orientations directly from the more easily extracted shape information, using a statistical predictive model constructed from a training sample of *ex vivo* DTI datasets. However, the efficiency of such statistics-based methods for the simulation of electrophysiology is yet to be measured. Consequently, the goals of this paper are two-fold: firstly, to build and apply a predictive model of fiber orientation within an EP simulation framework, and secondly to perform a comparative study between the DTI-, the existing Streeter-, and the statistics-based fiber models in order to quantify the potential improvement in the accuracy of EP simulations.

2 Methods

2.1 DTI-Based Fibers

The first part of this study consists of producing DTI-based models of fiber orientations, which is achieved based on a publicly available database of *ex vivo* DTI data from John Hopkins University [2]. The sample includes nine DTI datasets, of which seven are normal cases and two failing hearts (which do not have infarct scars). To obtain the DTI-based fiber models for these subjects, we first extract the myocardial surfaces by using a modified active shape model (ASM) search [13]. We use the maps of fractional anisotropy (FA) to search for the myocardial edges (one can also use the sharper B0 maps instead of the FA maps, in our case we perform a manual correction of the segmentations to ensure global, as well as local, accuracy in the boundary definitions). The obtained myocardial surfaces are then used to extract a mean volumetric mesh using the well-known TetGen tool, which is subsequently propagated to all subjects using Thin Plate Spline (TPS). We finally calculate the fiber orientation vectors at all myocardial locations of the volumetric meshes by using linear interpolation based on the 8 nearest voxels for each node (thus with minimal smoothing). Note that the fiber orientation at each voxel is taken to be equal to the first eigenvector of the diffusion tensor as obtained from the DT-MRI datasets.

2.2 Streeter Fiber Model

To assess the relative performance of the proposed predictive model for EP simulation, we also implement the well-known synthetic fiber model introduced by Streeter [1]. It consists of set of a rules which describe the distribution of the myocytes in the myocardium as observed initially in histological studies [1] and later on confirmed with *ex vivo* image data [2]. Essentially, it was shown that the longitudinal fiber direction rotates clockwise thorough the ventricular walls in parallel lines, while varying smoothly the elevation angle α between the fiber orientations and the short-axis planes, from the endocardium ($+\alpha_0$) to the epicardium ($-\alpha_0$). To derive a mathematical formulation of the elevation angle, one must first calculate for each node position $\mathbf{x}$ the normalized distance to the walls, *i.e.*,

$$e(\mathbf{x}) = \frac{d_{endo}(\mathbf{x})}{d_{endo}(\mathbf{x}) + d_{epi}(\mathbf{x})}. \tag{1}$$

The challenge here is in calculating the minimal distances from each node to the myocardial walls, which can be computationally demanding depending on the method (e.g., Laplacian or nearest neighbor approach). Finally, the variation in the elevation angle can be derived as follows [7]:

$$\alpha(\mathbf{x}) = \alpha_0 (1 - 2e(\mathbf{x}))^n, \tag{2}$$

where $n = 1$ for linear or $n = 3$ for cubic interpolation between the endocardial and epicardial walls. For the canine datasets α_0 is typically set to 60 degrees [14].

2.3   Predictive Fiber Model

In this section we describe the implemented predictive fiber model as developed by Lekadir *et al.* [12], which is applied in this paper for statistics-based EP simulation. The fundamental goal of the approach is to statistically encode the relationship between myocardial morphology and fiber, with the ultimate aim to predict the missing fiber information from the more easily extracted subject-specific shape. This can be done for example by using the DTI-based fiber models extracted in Section 2.1 as a training population.

The implementation of the approach requires two key stages. Firstly, to enable a consistent manipulation of the training data within the proposed predictive model, it is important to use a suitable representation of the fibers. In particular, fiber orientations by definition should be non-directional, which means opposite vectors $\mathbf{y}$ and $-\mathbf{y}$ must correspond to the same fiber. The solution to this problem consists of a Knutsson mapping from a 3D fiber vector $\mathbf{y}$ to a 5D representation $M_K : R^3 \rightarrow R^5$ such that $M_K(\mathbf{y}) = M_K(-\mathbf{y})$. Given a unit vector characterized by the angles θ, φ of its spherical representation, we obtain the following fiber coordinates:

$$\left(\sin^2\theta\cos 2\phi, \sin^2\theta\sin 2\phi, \sin 2\theta\cos\phi, \sin 2\theta\sin\phi, \sqrt{3}\left(\cos^2\theta - \frac{1}{3}\right)\right). \tag{3}$$

Once the input shapes $\mathbf{X}$ and output Knutsson fiber coordinates $\mathbf{Y}$ are obtained, the second stage involves the estimation of an optimal regression model, which is done using kernel partial least squares regression (PLSR) [15], *i.e.*, new predictions can be written in the form:

$$\hat{\mathbf{y}} = \mathbf{K}_t \mathbf{U}(\mathbf{T}^T \mathbf{K}\mathbf{U})^{-1}\mathbf{T}^T\mathbf{Y}, \tag{4}$$

where **K** refers the kernel Gramm matrix and **T** and **U** are the regression matrices (see [12], [15] for more details). The advantage of this regression technique is that it extracts in the input space (myocardial shapes) the directions that are optimal for the prediction of the output (fibers). Also, while the Streeter model uses predefined shape variables (*e.g.*, distances to myocardial walls, ventricular axis) for all fiber nodes, the predictive technique extracts the relevant shape predictors specifically for each myocardial node based on statistical criteria.

It is important to note that the predictive fiber model requires point correspondence between the ventricular meshes. This is achieved in this work by using key anatomical landmarks, which include the location of the apex, mitral valves and LV/RV junction points.

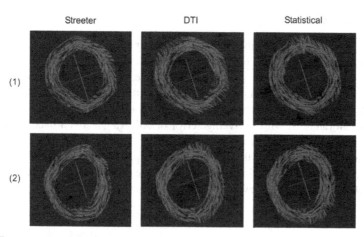

Fig. 1. Two cross-section examples of myocardial fiber orientations as obtained with the Streeter, DTI- and predictive models

2.4   Electrophysiological Simulation

An important goal of the fiber personalization is ultimately to simulate with the highest accuracy the electrical activation, both in terms of propagation times and patterns. Currently, the most popular approach to achieve this is the Eikonal equation, which has been used and validated extensively for the description of wave front propagation [6], [8]. Existing research has shown that a limited loss of accuracy can be obtained with such models, while obtaining computationally efficient solutions, thus allowing for large scale EP simulations to be performed [3].

In this paper, we model myocardial electrophysiology using the anisotropic formulation by Pashaei et al. [8]. One important advantage of the method is that it couples the cardiac conduction system to cardiac myocytes through a model of Purkinje-ventricular junctions, which leads to a more realistic electrical activation of the ventricles. The EP wave propagation is modeled only through element edges in both the myocardium and the cardiac conduction system (including the Purkinje tree) based on a one-manifold implementation of the fast marching method (FMM).

Information exchange across these two domains is achieved through their nodal connectivity thus keeping the solution on a one-manifold.

For consistent comparisons, the same EP parameters were used for all simulations (the conduction velocity for the Purkinje system was set to 3.0 m/s, the working myocardial velocity along the fiber orientations to 0.75 m/s, and the transversal conduction velocity to 0.3 m/s, as reported in the literature [14]). For the personalization of the fiber orientations, the simulated dataset was removed from the training of the predictive fiber model (leave-one-out).

3   Results

3.1   Normal Cases

Examples of the myocardial fiber orientations obtained using the different models are given in Fig. 1. For numerical evaluation, the total activation times for all simulations were calculated. The Streeter vs. DTI and predictive vs. DTI errors (in %) are plotted in Fig. 2 for all normal subjects. It can be seen that the statistical fiber personalization derives low errors in the estimation of the activation times in a consistent manner throughout all datasets, with an average error of only 1.98% and a maximum error of 4.28%. For two cases, the error is less than 0.5%, which is a significant accuracy for EP simulation given the fact that the training sample did not include those datasets (leave-one-out experiments). In contrast, the average error for the Streeter model is equal to 10.4%, and the error even reaches 15% for subject 1.

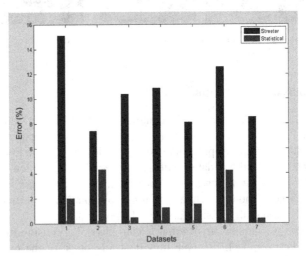

Fig. 2. Error bars comparing the Streeter and predictive fiber models for the estimation of total activation times in the normal datasets

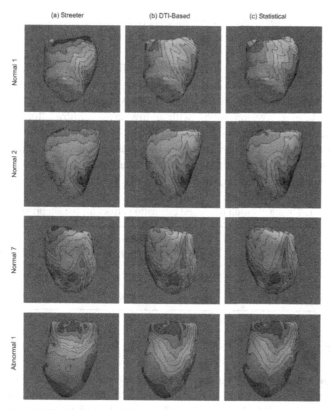

Fig. 3. Examples of EP simulations obtained with the different fiber models, showing good agreement between the DTI- and statistics-based results

Fig. 3 displays three examples of the obtained activation patterns on the ventricular surfaces (normal subjects 1, 2, and 7). For each subject, the longest of the three total activation times was used to determine a set of ten equally spaced isochrones, from which it can be observed that the simulated propagation patterns using the Streeter model (a) can vary from those obtained with the original DTI data (b). In contrast, the statistical approach approximates well the DTI-based electrical propagations for all subjects, both globally and locally. The obtained EP simulations with the DTI- (b) and the statistics-based fiber structures (c) are nearly identical for all cases.

3.2   Abnormal Cases

In this section we assess the ability of the predictive model constructed using the seven normal datasets to extrapolate the fiber modeling for abnormal cases. To achieve this, the EP simulation was applied to two failing canine hearts based on the DTI-, Streeter, and predictive fiber models. The total activation times and associated errors are summarized in Table 1, where it can be seen that the predictive fiber model enables to obtain accurate results for both cases, with an average error of less than

1%. On the contrary, the Streeter fiber model produces one accurate estimation of the activation times (subject 1) but introduces significant error (13.64%) for the second failing case (see Table 1).

Fig. 3 (last row) shows the activation patterns for the abnormal heart 1, where it can be seen that the maps obtained with the DTI- (b) and statistics-based (c) fiber models are almost identical. On the other hand, while the synthetic Streeter model estimates the total activation times with a low error (2.66%) for subject 1, the associated electrical propagation curves in Fig. 3 differ from those obtained with the DTI fiber model (b). This demonstrates the importance of accurate personalization of fibers, not only for the estimation of the total activation times but also to realistically simulate the electrical activation patterns both globally and locally.

Table 1. Total activation times and estimation errors for the abnormal cases

	TAT (ms) Streeter	TAT (ms) DTI	TAT (ms) Statistical	Error (%) Streeter-DTI	Error (%) Stat.-DTI
Abnormal 1	74.79	**72.85**	**73.51**	2.66	**0.90**
Abnormal 2	88.97	**78.29**	**78.79**	13.64	**0.63**

4   Conclusions

This paper presented a study on the application of statistics-based fiber models for electrophysiological simulation, with comparison to the widely used Streeter model and the more realistic DTI-based fibers. These initial results clearly promote the use of statistics-based fiber models, as they significantly improve the estimation accuracy of the electrical activation times (from an average of 10% to 1% errors) and propagation patterns. This performance is due to the fact that the predicted fibers, unlike the synthetic Streeter model, are more realistic and conform better to the subject-specific distribution of the myocytes.

Several improvements can be made to the current implementation of the predictive fiber model, for example by using tensor descriptors instead of the Knutsson mapping, which would allow keeping the orthotropic information. Furthermore, a more detailed comparison to an average fiber atlas as the one developed by Peyrat et al. [11] will be performed as a future work in order to quantify in detail the benefits of linking subject-specific morphology to fiber structure.

Acknowledgments. This work was funded partly by the euHEART project from the FP7 of the European Union, and partly by the CDTI CENIT-cvREMOD grant of the Spanish Ministry of Science and Innovation. Karim Lekadir was supported by a Juan de la Cierva research fellowship from the Spanish Ministry of Science and Innovation.

References

1. Streeter, D.D., Spotnitz, H.M., Patel, D.P., Ross, J., Sonnenblick, E.H.: Fiber orientation in the canine left ventricle during diastole and systole. Circulation Research 24(3), 339–347 (1969)

2. Helm, P., Beg, M.F., Miller, M.I., Winslow, R.L.: Measuring and mapping cardiac fiber and laminar architecture using diffusion tensor MR imaging. Annals of the New York Academy of Sciences 1047(1), 296–307 (2005)

3. Wallman, M., Smith, N.P., Rodriguez, B.: A comparative study of graph-based, eikonal, and monodomain simulations for the estimation of cardiac activation times. IEEE Transactions on Biomedical Engineering 59(6), 1739–1748 (2012)

4. Sebastian, R., Zimmerman, V., Romero, D., Sanchez-Quintana, D., Frangi, A.: Characterization and modeling of the peripheral cardiac conduction system. IEEE Transactions on Medical Imaging 32(1), 45–55 (2013)

5. Pitt-Francis, J., Pathmanathan, P., Bernabeu, M.O., Bordas, R., Cooper, J., Fletcher, A.G., Mirams, G.R., Murray, P., et al.: Chaste: A test-driven approach to software development for biological modelling. Computer Physics Communications 180(12), 2452–2471 (2009)

6. Sermesant, M., Konukoglu, E., Delingette, H., Coudière, Y., Chinchapatnam, P., Rhode, K.S., Razavi, R., Ayache, N.: An anisotropic multi-front fast marching method for real-time simulation of cardiac electrophysiology. In: Sachse, F.B., Seemann, G. (eds.) FIHM 2007. LNCS, vol. 4466, pp. 160–169. Springer, Heidelberg (2007)

7. Potse, M., Dubé, B., Richer, J., Vinet, A., Gulrajani, R.M.: A comparison of monodomain and bidomain reaction-diffusion models for action potential propagation in the human heart. IEEE Transactions on Biomedical Engineering 53(12), 2425–2435 (2006)

8. Pashaei, A., Romero, D., Sebastian, R., Camara, O., Frangi, A.F.: Fast multiscale modelling of cardiac electrophysiology including Purkinje system. IEEE Transactions on Biomedical Engineering 58(10), 2956–2960 (2011)

9. Bayer, J.D., Blake, R.C., Plank, G., Trayanova, N.A.: A novel rule-based algorithm for assigning myocardial fiber orientation to computational heart models. Annals of Biomedical Engineering, 1–12 (2012)

10. Weese, J., Groth, A., Nickisch, H., Barschdorf, H., Weber, F., Velut, J., Castro, M., Toumoulin, C., et al.: Generating anatomical models of the heart and the aorta from medical images for personalized physiological simulations. Medical & Biological Engineering & Computing, 1–11 (2013)

11. Peyrat, J.-M., Sermesant, M., Pennec, X., Delingette, H., Xu, C., McVeigh, E.R., Ayache, N.: A computational framework for the statistical analysis of cardiac diffusion tensors: Application to a small database of canine hearts. IEEE Transactions on Medical Imaging 26(11), 1500–1514 (2007)

12. Lekadir, K., Ghafaryasl, B., Muñoz-Moreno, E., Butakoff, C., Hoogendoorn, C., Frangi, A.F.: Predictive modeling of cardiac fiber orientation using the Knutsson mapping. In: Fichtinger, G., Martel, A., Peters, T. (eds.) MICCAI 2011, Part II. LNCS, vol. 6892, pp. 50–57. Springer, Heidelberg (2011)

13. Cootes, T.F., Cooper, D., Taylor, C.J., Graham, J.: Active shape models - Their training and application. Computer Vision and Image Understanding (CVIU) 61(1), 38–59 (1995)

14. Kerckhoffs, R.C., Faris, O.P., Bovendeerd, P.H., Prinzen, F.W., Smits, K., McVeigh, E.R., Arts, T.: Timing of depolarization and contraction in the paced canine left ventricle. Journal of Cardiovascular Electrophysiology 14(s10), S188-S195 (2003)

15. Rosipal, R., Trejo, L.J.: Kernel partial least squares regression in reproducing kernel Hilbert space. The Journal of Machine Learning Research 2, 97–123 (2002)

Automatic Extraction of the 3D Left Ventricular Diastolic Transmitral Vortex Ring from 3D Whole-Heart Phase Contrast MRI Using Laplace-Beltrami Signatures

Mohammed S.M. ElBaz[1], Boudewijn P.F. Lelieveldt[1,2], Jos J.M. Westenberg[1], and Rob J. van der Geest[1]

[1] Division of Image Processing, Department of Radiology, Leiden University Medical Center, Leiden, The Netherlands
[2] Department of Intelligent Systems, Delft University of Technology, Delft, The Netherlands
m.s.m.m.el_baz@lumc.nl

Abstract. In this work, a new method is proposed for automatic extraction of the left ventricular diastolic transmitral vortex ring from 3D whole-heart three directional Phase Contrast MRI. The proposed method consists of two parts, training and extraction. In the training step, an average reference signature of the complex transmitral vortex ring is captured from training subjects using Laplace-Beltrami spectrum and the Lambda2 method. In the vortex extraction step, the trained signature is used to identify the vortex ring by performing an iterative search for the vortex object with minimum distance from the trained signature. The proposed method is validated on a dataset of 8 healthy volunteers with 32 observed diastolic vortex rings. The method was able to successfully extract 27 diastolic vortex rings from a total of 32. Furthermore, the conducted experiments showed the capability of the proposed method in dealing with vortex shape changes that occur between the phases of early and late diastolic filling.

1   Introduction

Vortex formation in intra-cardiac flow patterns has recently gained much interest due to its vital role in keeping balance between blood motion and stresses of surrounding structures. Vortices are complex flow structures that evolve as a result of a change in velocity direction around an imaginary axis. In the cardiac Left Ventricle (LV), during early and late diastolic filling, the flow behind the mitral valve develops as a closed vortex tube: a vortex ring [1]. Vortex rings are frequently observed in nature because of their stability [1]. Recent studies have shown that transmitral vortex rings evolve in the LV during rapid early filling (E-wave) and late filling (A-wave) [11,13]. These vortex rings help in improving blood transport through the ventricle towards the aorta, minimizing the loss of energy and preventing blood stagnation [1,6,13]. Moreover, patients with diastolic dysfunction have been shown to form different diastolic vortex rings compared to healthy volunteers [5,10,12]. This makes vortex ring analysis a promising tool for detection of diastolic blood flow abnormalities. Nevertheless, most of the reported studies are based on Computational Fluid Dynamics (CFD) simulations [1,6] or Echocardiography [5,10]. CFD simulations usually require simplifications of the anatomy (i.e. cardiac chambers) or boundary conditions, which

O. Camara et al. (Eds.): STACOM 2013, LNCS 8330, pp. 204–211, 2014.
© Springer-Verlag Berlin Heidelberg 2014

might result in simulated blood flow velocities different from the actual flow. In echocardiography, generally only one single velocity component out of the three velocity components can be acquired providing limited flow velocity information.

Phase Contrast MRI (PC-MRI), also referred to as Velocity-Encoded MRI, can acquire all the three directional velocity components (in-plane and through-plane) of the blood flow relative to the three spatial dimensions and over the cardiac cycle, providing a powerful tool for cardiac flow analysis. In [12], Toger et al. used PC-MRI flow data to measure diastolic vortex ring volume using manual delineation of the vortex ring boundary from visualized Lagrangian coherent structures. They used the measured vortex volumes to differentiate between healthy volunteers and patients with dilated ischemic cardiomyopathy. In [4], Eriksson et al. proposed to quantify the intraventricular cardiac blood flow based on the visualization of PC-MRI data using pathline extraction, which allowed them to subdivide the intracardiac flow into four components based on their rates of passage relative to the cardiac cycle. In [2,3], flow visualization techniques (e.g. particle tracing, stream lines, streaklines,...etc) for PC-MRI flow were used to qualitatively assess the aorta function. Nevertheless, in most of these studies, vortex rings were defined qualitatively using flow visualization techniques (e.g. as region of swirling pathlines or steamlines), which might suffer from observer bias or high cluttered data. In [14], ElBaz et al. used the lambda2 method which is a quantitative method to define vortex rings. However, vortex rings were then extracted manually which is a tedious and time consuming process.

Due to the complex intra-cardiac blood flow, vortex rings are neither the only nor always the largest vortex object in the heart. Thus, using simple metrics (e.g. vortex size or location) is not enough to extract the LV vortex ring from surrounding vortex structures, similar in size, close in space, but different in shape. Furthermore, cardiac vortex rings are not ideally shaped rings but rather complex structures that tend to have a quasi-ring-like shape (Fig.1). All these factors make automatic vortex ring extraction from PC-MRI flow data a difficult and challenging task.

In this paper, we propose a novel method for automatic extraction of diastolic transmitral vortex rings from three-directional, three dimensional time resolved Phase Contrast MRI flow data during the rapid early (**E**) and late (**A**) filling phases. In the proposed work, we use a cardiac-vortex-specific shape signature to tackle the complex cardiac vortex shape and structure problems.

The proposed method consists of two parts. First, vortex structures are identified from the PC-MRI flow field using the Lambda2 method [7]. From this, a cardiac vortex ring signature is defined using the Laplace-Beltrami spectrum method [8]. Second, the cardiac vortex is extracted from the PC-MRI flow field by searching iteratively for the object with the best signature match relative to the reference signature. To the best of our knowledge, this work is the first attempt to extract vortex rings automatically from Phase Contrast MRI flow data in general and from the LV in particular.

2 Methodology

2.1 Vortex Identification Using the Lambda2 Method

The first step towards vortex ring extraction is to identify vortex structures from the MRI flow field. To achieve this, we use the Lambda2 method [8] to detect vortex cores as it is considered the most accepted vortex identification technique [1]. Furthermore, the Lamda2 method is a quantitative detection method, i.e. it does not depend on visualization techniques but rather on the physical fluid dynamics definition of the vortex structure. The input for the Lambda2 method are the three velocity components of the velocity vector field. Let U, V and W denote the three velocity components of the flow field acquired using PC-MRI and X, Y, Z denote the three spatial dimensions. Then the Lambda2 method can be applied as follows. First, the velocity gradient tensor $\mathbf{J}$ is computed as

$$\mathbf{J} = \begin{bmatrix} \frac{\partial u}{\partial x} & \frac{\partial u}{\partial y} & \frac{\partial u}{\partial z} \\ \frac{\partial v}{\partial x} & \frac{\partial v}{\partial y} & \frac{\partial v}{\partial z} \\ \frac{\partial w}{\partial x} & \frac{\partial w}{\partial y} & \frac{\partial w}{\partial z} \end{bmatrix} \tag{1}$$

Second, the tensor $\mathbf{J}$ is decomposed into its symmetric part, the strain deformation tensor $\mathbf{S} = \frac{\mathbf{J}+\mathbf{J}^T}{2}$ and the antisymmetric part, the spin tensor $\boldsymbol{\Omega} = \frac{\mathbf{J}-\mathbf{J}^T}{2}$, where T is the transpose operation. Then, eigenvalue analysis is applied only on $\mathbf{S}^2 + \boldsymbol{\Omega}^2$. Finally, a voxel is labeled as part of a vortex core only if it has two negative eigenvalues i.e. if $\lambda_1, \lambda_2, \lambda_3$ are the eigenvalues whereas $\lambda_1 \geq \lambda_2 \geq \lambda_3$ then a voxel is labeled as vortex core if its $\lambda_2 < 0$. However, usually the velocity data is noisy, and as a result of which $\lambda_2 < 0$ gives cluttered results. Therefore, a λ_2 threshold, $T_{\lambda_2} < 0$, is applied instead to allow separation of strong vortex structures from weaker ones. Using the detected vortex voxels, a vortex structure is defined as connected region of these voxels. In this work, we used connected component analysis (CCA) [9] to define the connected vortex cores. The CCA performance is governed by the threshold T_{λ_2} i.e. it is important that T_{λ_2} results in separate vortex structures for CCA to be able to define them as seperate objects. The Lambda2 method yields the vortex structures in the flow field. These vortex structures are usually visualized as isosurfaces with T_{λ_2} as the isovalue.

It is important to note that the Lambda2 method detects all vortex structures from the flow field i.e. vortex rings may be included in the extracted vortices but not all extracted vortices are vortex rings. The output of this step is converted to isosurfaces of the detected vortex structures from the PC-MRI flow field data (Fig.1). The vortex shape signature is subsequently captured by applying the Laplace-Beltrami method to these isosurface meshes.

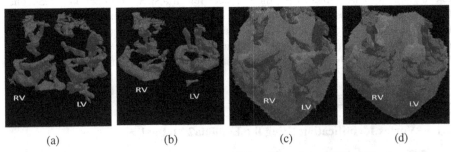

(a)   (b)   (c)   (d)

Fig. 1. (a,b) Lambda2 isosurfaces of peak early filling (**E**) and late filling (**A**) phases respectively (c,d) their respective heart position in whole heart region of interest magnitude images

2.2   Capturing Vortex Ring Shape Signature Using Laplace-Beltrami Spectrum

From Fig.1, it is obvious that cardiac vortex rings are rather complex structures which tend to have a quasi-ring-like shape. Therefore, a method for extraction of cardiac vortex rings should capture the features specific for cardiac vortex rings. We achieve this by using the recently introduced Laplace-Beltrami spectral shape signature [8]. This spectral shape signature is a global shape signature computed only from the object's inherent geometry (e.g. curvature, surface area and volume). Furthermore, this signature can be used to compare objects independent of their representation, position and size. This signature is defined as the beginning sequence of the Laplace-Beltrami (LB) differential operator. That is, for a given manifold M, if the LB operator is denoted by Δ, then the Laplacian eigenvalue equation can be written as :

$$\Delta f = -\lambda f \qquad (2)$$

where λ is a real scalar value corresponding to the eigenvalue of the Laplacian Δ and f corresponds to its eigenvectors. The shape spectral signature is then defined as the diverging sequence of eigenvalues $0 < \lambda_1 \leq \lambda_2 \leq \lambda_3 \leq \cdots + \infty$. This spectrum is truncated at the dth eigenvalue where d is application specific, and determined empirically. In our case, we apply the LB operator on the Lambda2 vortex isosurfaces which are discrete triangle meshes, hence, we solve (2) using a finite element method and apply the discrete Laplace-Beltrami (LB) operator and follow the same procedure as described in [8] to capture the LB spectrum for the vortex isosurface.

Though similar, cardiac vortex rings differ between subjects. Therefore, we derive an averaged signature from multiple subjects using Laplace-Beltrami analysis as follows. First, for each training subject, the peak early filling (**E** phase) transmitral vortex ring isosurface is manually selected from the identified vortex structures. Second, for each extracted vortex, the Laplace-Beltrami signature is captured as described above. Then, every signature is normalized by both slope of its fitting line and the volume of the vortex isosurface (i.e. the number of voxels in the isosurface) [10]. The reason for this normalization is to make signatures scale invariant. Finally, signature average is computed. Through the rest of the paper we denote the computed vortex shape signature average by **VS.** Due to the representation, position and size invariance properties of the LB signature [8], no shape registration is required prior to averaging. The steps for the vortex ring signature extraction from one subject are illustrated in Fig.2.

Of note, in addition to the E-phase averaged signature, we tested a signature trained on shapes of both phases (**E** and **A**) vortex rings. However, this provided identical results as for using only **E**-phase averaged signature.

2.3   Vortex Ring Extraction

The vortex ring extraction starts by identifying the vortex structures from the PC-MRI data using Lambda2 method as explained in section 2.1 Then, the normalized signature of each vortex object in the desired frame is captured using Laplace-Beltrami spectral shape analysis as explained in the previous section. For each vortex object in the current frame, its signature distance D_{sig} from the reference signature **VS** is computed as the L2 norm and computed as:

$$D_{sig}(m) = \sum(g_m - \textbf{VS})^2 \,, \, m=1\ldots M \qquad (3)$$

with g_m being the mth object signature and M the total number of vortex objects in the frame under processing. The extracted vortex ring is then defined as the vortex structure with the minimum D_{sig}.

3   Experiments

3.1   Data and Preprocessing

The proposed method was evaluated on a data from eight healthy volunteers (mean age: 40±15 years) who underwent three-dimensional (3D), time resolved, three-directional Phase Contrast (VE) MR imaging at 1.5 T (Philips). VE MRI was performed in a 3D isotropic dataset of 4.2×4.2×4.2mm^3 covering all 4 cardiac chambers. Retrospective gating with 30 phases with average temporal resolution of 30 ms were reconstructed and velocity sensitivity of 150cm/s in all directions were used. This data was then linearly interpolated spatially to result in a 1 mm^3 spatial resolution. The whole heart (not just the LV) region was then outlined manually from all slices and time frames. There are two reasons behind segmenting the whole heart region instead of just the LV. First, to investigate the ability of our method in extracting the LV vortex rings in the presence of other vortex structures formed in other ventricles. Second, to avoid the need for LV segmentation from the PC-MRI magnitude images (Fig.1. b and d) which usually suffer from low contrast between LV and right ventricle (RV) boundaries making LV segmentation a difficult task.

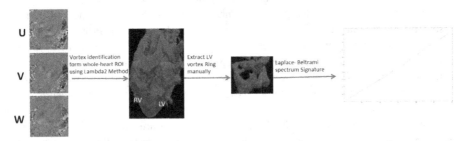

Fig. 2. Steps of the proposed vortex ring shape signature extraction from one subject, U,V,W are three volumes representing the PC-MRI flow field velocity components

3.2   Diastolic Vortex Ring Extraction

Using the manually segmented whole heart flow field volumes resulting from the previous step, vortex structures were identified using the Lamda2 method. After Applying threshold T_{λ_2} (as explained in Sec. 2.1), connected component analysis method (CCA) [9] was then applied to define the identified vortices as connected vortex objects. After that, LV vortex rings were labeled manually to be used as ground truth. In this work, for each subject, two observed rings were labeled from each of the rapid early (E) and late filling (A) diastolic phases. The two early filling rings correspond to the rings of the peak early filling PC-MRI phase and the subsequent frame. Similarly, the two late filling rings were labeled from the peak late filling phase and the subsequent frame. These were the frames in which vortex rings were observed consistently in all 8 subjects. From the eight volunteers, in total 32 LV vortex rings were manually labeled which then used as the ground truth to evaluate the proposed extraction method. For computing the Laplace-Beltrami (LB) signature [8], the vortex shape signature is captured from the Lambda2 isosurfaces with T_{λ_2} as isovalue.

To quantitatively evaluate the proposed method and to avoid bias in the selection of the average signature **VS**, a leave-one-out cross-validation approach was used. The average signature **VS** was computed from 7 subjects out of the available 8 subjects (i.e. computed as average of the corresponding 28 vortex signatures). This **VS** is then used to extract the LV vortex rings from the 4 aforementioned frames of the left out subject. This is repeated 8 times, leaving out different subjects. To evaluate the extraction performance we used the precision criterion, which was computed as the proportion TP/(TP+FP) where TP stands for the true positive i.e. the number of correctly extracted LV vortex rings, FP for false positive i.e. the number of the mis-extracted LV vortex rings.

Parameter Selection

In the proposed method there are two empirically determined parameters, T_{λ_2} and d. T_{λ_2} is application and subject specific. In this study, T_{λ_2} was manually adjusted per subject until meaningful vortex rings could be differentiated from surrounding structures. In our experiments, T_{λ_2} in the range of [2-5] μ (with μ as the λ_2 average of voxels with $\lambda_2 < 0$) was found to give good results. Second, in the applied Laplace-Beltrami analysis, a signature of 300 eigenvalues (i.e. d=300) was sufficient in all experiments.

4   Results

The overall precision is 0.844, detailed results for the performance over the two diastolic phases are given in Table 1, where every phase has a total of 16 LV transmitral vortex rings to be extracted. In the reported results, vortex rings were extracted from an average of 43 different sized surrounding vortex structures in the E-phases and an average of 30 structures in the A-diastolic phases. The proposed method failed in extracting only 5 rings, 1 from the E phase and 4 from A, out of the total 32 vortex rings.

Table 1. LV Transmitral vortex ring extraction results

Phase	E (n=16)	A (n=16)	Total (n=32)
True Positive	15	12	27
False Positive	1	4	5

5   Discussion and Conclusion

Our results show that the proposed cardiac-vortex-specific signature based extraction is rather accurate in extracting LV diastolic transmitral vortex rings from whole heart PC-MRI with 27 successfully extracted LV vortex rings out of the total 32 rings yielding an overall precision of 0.844. In all 5 failed cases, the proposed algorithm extracted the RV C-shape or incomplete rings instead of the LV ring i.e. it was successful in ring extraction but could not differentiate between the RV partial rings and the more complete LV vortex rings. This could be due to the similarity in shape (e.g. curvature and complexity) of RV and LV vortex rings. Moreover, in all failed experiments, the LV vortex ring was ranked second after the RV partial ring based on the distance defined in Eqn.3 with a small difference of 0.16 ± 0.23 from the highest rank while the third ranking structure (not ring) was more distant (2.72 ± 1.90) from the highest ranking structure. It is important to note that the proposed E-phase trained average signature was able to detect most of the A-phase rings (12 out of 16), which shows the ability of the proposed method to deal with shape variability of the transmitral vortex rings between the E and A diastolic phases. The proposed method is automatic relative to the LV vortex ring extraction process. In this work, the whole heart region was still segmented manually from PC-MRI as automatic segmentation is out of this paper's focus. On the other hand, vortex identification is a complex fluid dynamics topic and no definite rigorous vortex definition is yet reached. In this work, we used the Lambda2 method which is the most commonly accepted fluid dynamics definition of a vortex [1]. Nevertheless, this method requires definition of T_{λ_2} threshold for defining meaningful vortex structures. To the best of our knowledge, no objective method has been reached yet for defining T_{λ_2}. Currently, we are working on developing a method for objective definition of this threshold. For the LB signature normalization, we evaluated different normalizations as suggested in [8], however, the best normalization in our case was to normalize by both the signature's fitting line slope and the vortex volume.

To our knowledge, this is the first attempt to automatically extract transmitral vortex rings from PC-MRI in general and from the LV in particular. Our results show that the proposed method is a promising technique for left ventricular vortex ring extraction. Furthermore, the results show the capability of the proposed method dealing with the vortex ring shape differences between the two diastolic (E and A) phases. As such, this work can be seen as a first step towards a quantitative understanding of cardiac vortex structures, their evolution and physiological implications. In addition, the proposed method could be used for vortex ring analysis in CFD simulations.

Acknowledgement. This work is supported by Dutch Technology Foundation (STW): project number 11626.

References

1. Kheradvar, A., Pedrizzetti, G.: Vortex formation in the cardiovascular system. Springer (2012)
2. Markl, M., et al.: Time-resolved 3-dimensional velocity mapping in the thoracic aorta: visualization of 3-directional blood flow patterns in healthy volunteers and patients. J. Comput. Assist. Tomogr. 28(4), 459–468 (2004)
3. Morbiducci, U., et al.: In vivo quantification of helical blood flow in human aorta by time-resolved three-dimensional cine phase contrast magnetic resonance imaging. Ann. Biomed. Eng. 37(3), 516–531 (2009)
4. Eriksson, J., et al.: Semi-automatic quantification of 4D left ventricular blood flow. JCMR, 12(9) (2010)
5. Jiamsripong, P., et al.: Impact of acute moderate elevation in left ventricular afterload on diastolic transmitral flow efficiency: analysis by vortex formation time. J. Am. Soc. Echocardiogr. 22(4), 427–431 (2009)
6. Domenichini, F., Pedrizzetti, G., Baccani, B.: Three-dimensional filling flow into a model left ventricle. J. of Fluid Mech. 539, 179–198 (2005)
7. Jeong, J., Hussain, F.: On the identification of a vortex. J. of Fluid Mech. 285(69) (1995)
8. Reuter, M., Wolter, F.E., Peinecke, N.: Laplace–Beltrami spectra as 'Shape-DNA' of surfaces and solids. Computer-Aided Design 38(4), 342–366 (2006)
9. Haralick, R., Shapiro, L.: Algorithms in Computer and Robot Vision, vol. I, pp. 28–48. Addison-Wesley (1992)
10. Kheradvar, A., et al.: Assessment of transmitral vortex formation in patients with diastolic dysfunction. JASE 25(2), 220–227 (2012)
11. Le, T.B., Sotiropoulos, F.: On the three-dimensional vortical structure of early diastolic flow in a patient-specific left ventricle. Eur. J. Mech. B Fluids 35, 20–24 (2012)
12. Töger, J., et al.: Vortex ring formation in the left ventricle of the heart: analysis by 4D flow MRI and Lagrangian coherent structures. Ann. Biomed. Eng. 40(12), 2652–2662 (2012)
13. Charonko, J., et al.: Vortices Formed on the Mitral Valve Tips Aid Normal Left Ventricular Filling. Ann. Biomed. Eng., 1–13 (2013)
14. ElBaz, M.S., et al.: Quantification of diastolic vortex shape deformation in left ventricular filling from 4D flow MRI. JCMR, 15(suppl. 1), P79 (2013)

Direct Myocardial Strain Assessment
from Frequency Estimation in Tagging MRI

Hanne B. Kause[1], Olena G. Filatova[1], Remco Duits[2,3], L.C. Mark Bruurmijn[2],
Andrea Fuster[2,3], Jos J.M. Westenberg[4], Luc M.J. Florack[2,3],
and Hans C. van Assen[1]

[1] Department of Electrical Engineering
[2] Department of Biomedical Engineering
[3] Department of Mathematics & Computer Science
Eindhoven University of Technology, The Netherlands
www.iste.nl
[4] Department of Radiology, Leiden University Medical Center, The Netherlands
h.b.kause@tue.nl

Abstract. We propose a new method to analyse deformation of the cardiac left ventricular wall from tagging magnetic resonance images. The method exploits the fact that the time-dependent frequency covector field representing the tag pattern is tightly coupled to the myocardial deformation and not affected by tag fading. Deformation and strain tensor fields can be retrieved from local frequency estimates given at least n (independent) tagging sequences, where n denotes spatial dimension. Our method does not require knowledge of material motion or tag line extraction. We consider the conventional case of two tag directions, as well as the overdetermined case of four tag directions, which improves robustness. Additional scan time can be prevented by using one or two grid patterns consisting of multiple, simultaneously acquired tag directions. This concept is demonstrated on patient data. Tracking errors obtained for phantom data are smaller than those obtained by HARP, 0.32 ± 0.14 px versus 0.53 ± 0.07 px. Strain results for volunteers are compared with corresponding linearised strain fields derived from HARP.

Keywords: Tagging Magnetic Resonance Imaging, Myocardial Deformation, Myocardial Strain, Gabor Transform, Frequency Analysis.

1 Introduction

Cardiovascular disease (CVD) is globally the leading cause of death[1] with estimated 17.3 million deaths in 2008 (30% of all global deaths). Therefore it is important to develop methods for diagnosis and therapy assessment at early stages of the disease. In literature it has been reported that heart disease may

---

[1] World Health Organization, Fact sheet N° 317, March 2013.

O. Camara et al. (Eds.): STACOM 2013, LNCS 8330, pp. 212–219, 2014.
© Springer-Verlag Berlin Heidelberg 2014

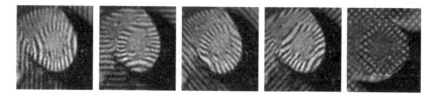

Fig. 1. Short-axis tMRI images of a left ventricle in systole

affect *strain* before remodeling occurs as a consequence of persistent heart dysfunction [1,2].

Speckle tracking echocardiography [3] is often the initial choice for cardiac movement assessment, since it is relatively cheap and widely accessible. However, the estimated strain depends on the angle of imaging and therefore the results are highly operator sensitive. In order to avoid this, some approaches employ tagging Magnetic Resonance Imaging (tMRI) [4,5]. tMRI uses spatial modulation of magnetisation (SPAMM) [5] to visualise the tissue deformation during the cardiac cycle by imprinting a *tagging* pattern in the tissue, see Fig. 1. The method that we present here is based on local frequency estimation and therefore insensitive to tag fading due to spin-lattice relaxation.

Different motion extraction methods have been developed to quantify cardiac function from tMRI. Harmonic Phase (HARP) [6,7] tracks material points based on the phase conservation principle and is the de facto standard for calculation of cardiac deformation. Other approaches are based on *local frequency* estimation, e.g. [8,9]. We pursue a new approach, but similar in spirit, in which the Gabor transform is employed to construct a local frequency representation of the tagging images. Subsequently, local frequency covector fields are extracted and used to determine the deformation tensor.

Since our method exploits frequency instead of amplitude information, it is more robust with respect to tag fading. Moreover, the method is designed to work with any number A of tag directions ($A \geq n$, where n denotes spatial dimension), which results in an overdetermined system of equations when $A > n$.

Thus, similarly to Qian et al. [8], we bypass the classic approach to assess strain through the gradient of the motion field [10]. In [8], line elements are considered, along which the deformation gradient is calculated and from which linearised radial and circumferential strains are obtained. These are thus confounded with shear strains. However, coordinate independence requires specification of the full strain tensor, which is the approach we choose here. The full deformation tensor thus obtained is used to disentangle radial, circumferential and shear components using the Lagrangian strain tensor. The first two are widely used measures for deformation. In this study, myocardial deformation is assessed using stripe tags in $A \in \{2, 4\}$ directions and grid tags, see Fig. 1.

2   Calculating Deformation from Local Frequency

Let us consider the tissue configuration at two distinct moments of time $t_0 = 0$ and $t > 0$. In an infinitesimally small neighbourhood, the global tagging pattern at t_0 can be considered as a constant frequency pattern ω_0. At time t this frequency pattern is deformed relative to the reference tissue configuration. The Gabor transform [11] offers a position-frequency representation of an image. In the continuous case this Gabor transform reads

$$G(\mathbf{p}, \omega) = \int_{\mathbb{R}^2} f(\mathbf{q})\overline{\psi(\mathbf{q} - \mathbf{p})}e^{-2\pi i(\mathbf{q}-\mathbf{p})\cdot\omega}d\mathbf{q}, \tag{1}$$

where $f : \mathbb{R}^2 \to \mathbb{R}$ is the 2-dimensional tagging image, $\psi : \mathbb{R}^2 \to \mathbb{C}$ the Gabor window, $\bar{\ }$ denotes complex conjugation and $\mathbf{p}, \omega \in \mathbb{R}^2$ are position and spatial frequency respectively. For our purpose, we extract a single frequency covector $\omega(\mathbf{p}(t), t)$ at each position $\mathbf{p}(t) = (x(t), y(t))$ at time t for each tag direction. A Gaussian window is chosen for ψ, for this has the best position-frequency localisation [12]. Details on the frequency selection method based on the Gabor transform can be found in [13].

While HARP and optical flow [10] assume phase conservation of a material point, this method employs phase difference constancy between tip and tail of a material vector inducing the covector transformation law

$$\omega_t = \omega_0 \mathbf{F}^{-1}, \tag{2}$$

where $\mathbf{F}$ is the deformation tensor [14] and row vectors ω_0, ω_t represent the local frequencies evaluated at corresponding material points. For more details see [15].

The relation between corresponding material points is not known, since we do not compute material motion. Hence, we assume that at the fiducial moment t_0, when the tagging pattern is applied in the scanner, the tag frequency is a known, global constant and therefore equal for every material point, obviating knowledge of material motion. We use this (unobserved) configuration at time t_0 as a reference.

Because a tMRI acquisition typically consists of at least two encoding directions, $A \geq 2$, Eq. (2) constitutes a system of equations that can be written as

$$\Omega_t = \Omega_0 \mathbf{F}^{-1}, \text{ with } \Omega = \begin{bmatrix} \omega^1 \\ \vdots \\ \omega^A \end{bmatrix}, \tag{3}$$

with upper indices $1, \ldots, A$ enumerating tag directions. The least squares solution for the deformation tensor at a material point at time t relative to t_0 can then be obtained via the pseudo-inverse

$$\mathbf{F} = \left(\Omega_t^{\mathsf{T}}\Omega_t\right)^{-1}\Omega_t^{\mathsf{T}}\Omega_0, \tag{4}$$

where T denotes transposed. The Lagrangian strain tensor is consequently defined as

$$\mathbf{E} = \frac{1}{2}(\mathbf{F}^{\mathsf{T}}\mathbf{F} - \mathbf{I}).\tag{5}$$

One can extract circumferential (E_{cc}), radial (E_{rr}) and shear (E_{cr}) strains

$$E_{cc} = \hat{\mathbf{e}}_c^{\mathsf{T}}\,\mathbf{E}\,\hat{\mathbf{e}}_c\;,\;\;E_{rr} = \hat{\mathbf{e}}_r^{\mathsf{T}}\,\mathbf{E}\,\hat{\mathbf{e}}_r\;,\;\;E_{cr} = \hat{\mathbf{e}}_c^{\mathsf{T}}\,\mathbf{E}\,\hat{\mathbf{e}}_r\tag{6}$$

with local unit radial and circumferential basis vectors $\hat{\mathbf{e}}_r$ and $\hat{\mathbf{e}}_c$.

The numerical implementation consists of the following steps:

1. Calculate the Gabor transform $G(\mathbf{p}(t), \boldsymbol{\omega}(t))$ in each pixel, Eq. (1), for all tag directions, using a Gaussian filter ψ with width σ.
2. Determine the local frequency $\boldsymbol{\omega}$ with highest intensity in the Gabor domain, excluding a priori unreachable areas limited by physical muscle deformation.
3. Compute the deformation tensor $\mathbf{F}$, Eq. (4) for every time t, relative to the reference time t_0.
4. Calculate the strain tensor $\mathbf{E}$, Eq. (5), and components E_{rr}, E_{cc} and E_{cr}, Eq. (6).

Here we used $\sigma = 4$. The optimal value of σ depends on the tag width.

3   Data Description

In this study, artificial, volunteer and patient tMRI data are used. The artificial tMRI data consisted of a series of 16 frames (64×64 pixels) of contracting and rotating rings, simulating the systolic phase in a single short-axis slice of the left ventricle, with two and four tag directions ($A \in \{2, 4\}$). The wall thickens and rotation decreases linearly with increasing radius, causing the endocardium to rotate more than the epicardium, inducing shear deformation. Tag fading was modelled by exponential decay and Rician noise was added.

For six volunteers, single short-axis slices of the left ventricle were obtained in 30 frames with SPAMM imaging, forming a whole heart cycle (systole and diastole). Data was acquired for four tag directions ($A = 4$) with a tag size of 7 mm. Volunteers were scanned with a 3T MRI scanner (Achieva, Philips Medical Systems, Best, The Netherlands) after informed consent and with permission given by the Medical Ethics Committee of the local institute.

A 2D multi-shot gradient-echo with Echo Planar Imaging with breath holding in end-expiration was used. Scan parameters were: TE 3.2 ms, TR 6.3 ms, flip angle 10°, slice thickness 10 mm and acquisition pixel size 1.34×1.34 mm^2 for volunteer 2 and 1.37×1.37 mm^2 for all other volunteers. Prospective triggering was used with a maximal number of reconstructed phases to ensure optimal temporal resolution. Contours of the myocardium were manually drawn.

The patient dataset was obtained with a 2D SPAMM gradient-echo sequence with breath holding in end-expiration. Scan parameters were: 1.5T MRI, TE 4 ms, TR 6.4 ms, a tag size of 7 mm, slice thickness 8 mm and acquisition pixel size 1.33×1.33 mm^2.

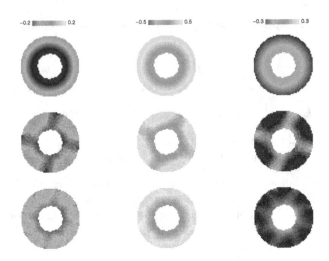

Fig. 2. Short-axis view of circumferential (left), radial (middle), and shear (right) components of the Lagrangian strain tensor for artificial data (mid-systole) comparing ground truth (top) with two diagonal (middle) and four (bottom) tag directions

Table 1. Average displacement errors of a rectangular grid for two and four tag directions using our novel method and HARP in mid-systole on artificial tMRI data

	hor. and vert. directions	diagonal directions	four directions
Gabor estimation	0.5 ± 0.22 px	0.38 ± 0.2 px	0.32 ± 0.14 px
HARP	0.54 ± 0.07 px	0.54 ± 0.05 px	0.53 ± 0.07 px

4   Results

For both artificial and volunteer data, local frequencies were calculated for all tag directions. We tested our method for both two ($A=2$) and four ($A=4$) tag directions. An adapted implementation was used for the grid tagging sequence of the patient dataset. In this case, two dominant peaks were located in the Gabor domain to determine the local frequencies corresponding to the two tag orientations of the grid. For all datasets, deformation tensors were computed according to Eq. (4). These were used to obtain radial, circumferential and shear strains, as presented in Fig. 2, Fig. 3 and Fig. 4.

For the artificial data, based on the deformation tensors from Eq. (4) displacements were calculated for a large number of points organised in a rectangular grid. The results of our method as well as deformed lattices calculated by our own implementation of HARP [16] are compared with the ground truth, using both four and two tag directions as inputs. We adapted HARP to accept four input image sequences by adding two equations to the iterative scheme and solving the obtained system with the least squares method. Although more precise implementations of HARP exist, we chose to use the faster implementation based on linearised strains [6]. As a consequence, these linearised strains can not be

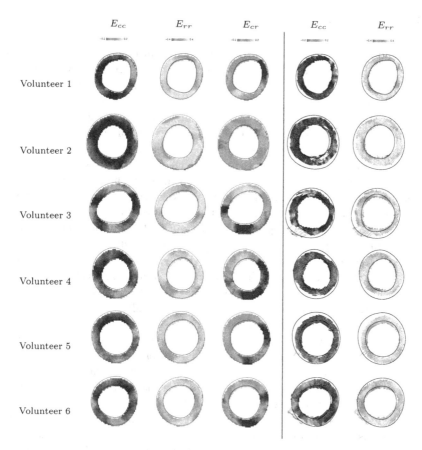

Fig. 3. Strain computed in short-axis view of the left ventricle (volunteer) for four tag directions. Left: Components of the Lagrangian strain tensor in end systole obtained with local frequency extraction. Right: Linearised strains obtained with HARP.

Fig. 4. Short-axis view of left ventricle *grid tagging* slice in end-systole (patient). Left: Circumferential, radial, and shear components of the Lagrangian strain tensor respectively. Right: Circumferential and radial linearised strains obtained with HARP.

directly compared with strain results of our method. Therefore, Table 1 shows the average displacement errors in pixels with respect to their true counterparts (average distance between calculated and true grids) in the mid-systolic frame using our novel method and HARP.

5   Discussion and Conclusion

We have shown that quantitative myocardial deformation can be accurately and robustly obtained from the Gabor frequency analysis of artificial tMRI data using four tag directions. Our method does neither require explicit knowledge of material motion nor tag line extraction. It is robust with respect to tag fading and can be straightforwardly generalised to any number of tag directions and to volumetric tagging data. Moreover, the proposed method uses the full form of the strain tensor instead of approximated linearised strain, which is used for HARP strain calculations. Due to the fact that estimated frequencies are directly used to calculate deformations, the smoothness of the frequency map has a major influence on the results. This effect is expected to be reduced by means of adapting the Gabor filter depending on the location in the muscle or muscle width, cf. [13], which is a subject for future work. Position dependent weighting in the least squares method Eq. (4) may lead to additional improvements. Nevertheless, subsequent approximation of tissue displacements performs comparably to HARP, while unlike with HARP using four tag directions improves the quality of the results in comparison with using two tag directions (cf. Table 1). Since HARP is an iterative method, it stops after a certain prescribed tolerance is achieved. This explains why the performance of HARP does not improve with more tag directions. Interestingly, and somewhat unexpectedly, a combination of diagonal tag patterns performs better than a horizontal-vertical tag pattern in artificial data. Considering the approximate rotational symmetry of the model, this is most likely caused by a discretisation effect related to the relative orientations of tag lines and pixel grid.

In current medical practice it is common to use only two tag directions for analysis. We used more directions to achieve more stable and homogeneous results, which may increase acquisition time. However, this can be prevented by using one or two grid patterns consisting of multiple, simultaneously acquired tag directions. Applicability of our method to a diagonal grid tagging sequence is shown on clinical data. Combining horizontal, vertical and diagonal tagging stripes in two grids will preserve robustness and keep acquisition times clinically acceptable.

Acknowledgements. The authors would like to thank dr. Brett Cowan and dr. Alistair Young, University of Auckland, New Zealand for providing the patient data set. This research is supported by the Dutch Technology Foundation STW, which is part of the Netherlands Organisation for Scientific Research (NWO), and which is partly funded by the Ministry of Economic Affairs.

References

1. Götte, M.J., van Rossum, A.C., Twisk, J.W.R., Kuijer, J.P.A., Marcus, J.M., Visser, C.A.: Quantification of regional contractile function after infarction: Strain analysis superior to wall thickening analysis in discriminating infarct from remote myocardium. Journal of the American College of Cardiology 37, 808–817 (2001)

2. Delhaas, T., Kotte, J., van der Toorn, A., Snoep, G., Prinzen, F.W., Arts, T.: Increase in left ventricular torsion-to-shortening ratio in children with valvular aorta stenosis. Magnetic Resonance in Medicine 51, 135–139 (2004)
3. Bohs, L., Geiman, B., Anderson, M., Gebhart, S., Trahey, G.: Speckle tracking for multi-dimensional flow estimation. Ultrasonics 38, 369–375 (2000)
4. Zerhouni, E.A., Parish, D.M., Rogers, W.J., Yang, A., Shapiro, E.P.: Human heart: Tagging with MR imaging—a method for noninvasive assessment of myocardial motion. Radiology 169(1), 59–63 (1988)
5. Axel, L., Dougherty, L.: MR imaging of motion with spatial modulation of magnetization. Radiology 171(3), 841–845 (1989)
6. Osman, N.F., Kerwin, W.S., McVeigh, E.R., Prince, J.L.: Cardiac motion tracking using CINE harmonic phase (HARP) magnetic resonance imaging. Magnetic Resonance in Medicine 42(6), 1048–1060 (1999)
7. Garot, J., Bluemke, D.A., Osman, N.F., Rochitte, C.E., McVeigh, E.R., Zerhouni, E.A., Prince, J.L., Lima, J.A.C.: Fast determination of regional myocardial strain fields from tagged cardiac images using harmonic phase MRI. Circulation 101(9), 981–988 (2000)
8. Qian, Z., Liu, Q., Metaxas, D., Axel, L.: Identifying regional cardiac abnormalities from myocardial strains using non-tracking-based strain estimation and spatio-temporal tensor analysis. IEEE Transactions on Medical Imaging 30(12), 2017–2029 (2011)
9. Arts, T., Prinzen, F.W., Delhaas, T., Milles, J.R., Rossi, A.C., Clarysse, P.: Mapping displacement and deformation of the heart with local sine-wave modeling. IEEE Transactions on Medical Imaging 29(5), 1114–1123 (2010)
10. Florack, L., van Assen, H.: A new methodology for multiscale myocardial deformation and strain analysis based on tagging MRI. International Journal of Biomedical Imaging, Article ID 341242 (2010), http://dx.doi.org/10.1155/2010/341242
11. Gabor, D.: Theory of communication. part 1: The analysis of information. Journal of the Institution of Electrical Engineers - Part III: Radio and Communication Engineering 93(26), 429–441 (1946)
12. Mallat, S.: A wavelet tour of signal processing. Academic Press (1999)
13. Duits, R., Führ, H., Janssen, B., Bruurmijn, M., Florack, L., van Assen, H.: Evolution equations on Gabor transforms and their applications. Applied and Computational Harmonic Analysis 35(3), 483–526 (2013), http://dx.doi.org/10.1016/j.acha.2012.11.007
14. Haupt, P.: Continuum Mechanics and Theory of Materials. Springer, Berlin (2002)
15. Bruurmijn, L., Kause, H., Filatova, O., Duits, R., Fuster, A., Florack, L., van Assen, H.: Myocardial deformation from local frequency estimation in tagging mri. In: Ourselin, S., Rueckert, D., Smith, N. (eds.) FIMH 2013. LNCS, vol. 7945, pp. 284–291. Springer, Heidelberg (2013)
16. Osman, N.F., McVeigh, E.R., Prince, J.L.: Imaging heart motion using harmonic phase MRI. IEEE Transactions on Medical Imaging 19(3), 186–202 (2000)

Estimation of Electrical Pathways Finding Minimal Cost Paths from Electro-Anatomical Mapping of the Left Ventricle

Rubén Cárdenes[1,2,*], Rafael Sebastian[3], David Soto-Iglesias[1], David Andreu[4],
Juan Fernández-Armenta[4], Bart Bijnens[1,5], Antonio Berruezo[4],
and Oscar Camara[1]

[1] PhySense, DTIC, Universitat Pompeu Fabra, Barcelona, Spain
[2] Fetal and Perinatal Medicine Research Group, IDIBAPS,
Hospital Clinic de Barcelona, Spain
[3] Computational Multi-scale Physiology Lab, Universitat de València, Spain
[4] Arrhythmia Section, Cardiology Department, Thorax Institute,
Hospital Clinic de Barcelona, Spain
[5] ICREA, Barcelona, Catalonia, Spain
ruben.cardenes@upf.edu

Abstract. The electrical activation of the heart is a complex physiological process that is essential for the understanding of several cardiac dysfunctions, such as ventricular tachycardia (VT). Nowadays, electro-anatomical mappings of patient-specific activation times on the left ventricle surface can be estimated, providing crucial information to the clinicians for guiding cardiac treatment. However, some electrical pathways of particular interest such as Purkinje or still viable conduction channels are difficult to interpret in these maps. We present here a novel method to find some of these electrical pathways using minimal cost paths computations on surface maps. Experiments to validate the proposed method have been carried out in simulated data, and also in clinical data, showing good performance on recovering the main characteristics of simulated Purkinje trees (e.g. end-terminals) and promising results on a real case of fascicular VT.

Keywords: electrical pathways, Purkinje, streamlines, fast marching, singular points, electro-anatomical mapping, cardiac arrhythmias, ventricular tachycardia.

1 Introduction

Ventricular tachycardia (VT) is one type of severe cardiac arrhythmias which is often treated with Radio-Frequency Ablation (RFA). The planning of these interventions has been substantially improved by integrating patient-specific imaging data with electro-anatomical mapping ([1,2]) to better targeting ablation

---

* Corresponding author.

sites. Nevertheless, some VTs are induced by patho-physiological mechanisms for which very limited patient-specific data is available such as abnormalities of the cardiac conduction system (CCS) in fascicular VTs.

The CCS is a heterogeneous network of cells responsible for the fast and coordinated distribution of the electrical impulses that triggers the contraction of the heart. In the ventricles, the CCS is composed of the His bundle (HB) and bundle branches (BB) that are connected to the most distal section, often called Purkinje (PK) system (see Fig. 1 left). The PK system plays a key role in the synchronous activation of the ventricles since it dictates the electrical activation sequence [3]. The CCS cannot be extracted from *in vivo* data due to the small size of its structures. Nevertheless, generic computational models have been developed capturing PK tree-like structure from *ex vivo* data available for different species, see [2] for a review of these models.

In severe fascicular VTs, the RFA intervention is based on the ablation of PK end-terminals (junctions between the CCS and the myocardial muscle) which are identified by manual detection of PK activations from electrocardiogram signals. Unfortunately, the detected PK end-terminals do not provide information about the whole tree-like structure of the CCS. We present here a methodology to obtain patient-specific activation lines from electro-anatomical maps based on finding geodesic paths. The methodology is a three-step procedure: first, the end terminals are detected from the electro-anatomical map; second, the electrical paths going from the detected end-terminals are reconstructed from the electro-anatomical map; and third, the conduction velocity of the geodesic paths are computed, to distinguish those that are close to the PK system from those that are only due to muscle activation.

For validation purposes we performed two experiments on synthetically generated electro-anatomical mappings. The first one is a simplistic simulation based on a surface fast marching solution to show the behavior of the geodesics generation. The performance of the detection of the end-terminals is evaluated with a second experiment where a simulated local activation map (LAT) is obtained for every point using detailed electrophysiological models and including a PK tree constructed with a L-systems-like method [2] on a LV geometry extracted from a CT image. Finally, we applied the algorithm to a real case of a patient with fascicular VT, finding relevant information about PK terminal distribution for the planning of ablation procedures.

2 Methods

2.1 Minimal Cost Paths

Minimal cost path computation is a well-known problem that is usually solved by Dijkstra algorithm [4] in graphs. For 2D/3D images, the pixels or voxels are used as the nodes and the distances between them are the weights associated to the edges of the graph. Computation of geodesics on surfaces has a higher level of complexity that has been addressed by other authors [5,6] in the past. In particular, given a surface $\mathcal{S}$, with a weighted map ψ defined on it, $\psi(\boldsymbol{x}) \; \forall \; \boldsymbol{x} \in \mathcal{S}$, and a surface point z denoted as the end point, the geodesic problem consists in finding the shortest weighted path $\Gamma_{\psi,y} : [0,1] \to S$ lying on the surface between any point on the surface y and z:

$$\Gamma_{\psi,y} = \underset{\gamma}{argmin}\{L_\psi(\gamma) : \gamma(0) = y, \gamma(1) = z\},$$

where $L_\psi(\gamma)$ is the curve weighted length

$$L_\psi(\gamma) = \int_0^1 \psi(\gamma(s))||\gamma(s)'||ds.$$

It is known [5,7] that a solution to this problem can be obtained using a hamilton-jacobi formulation, solving the eikonal equation for $\phi(x)$, that will define a distance map on the surface starting from z

$$||\nabla\phi(x)|| = 1/\psi(x), \ \forall \ x \in S,$$

with initial condition $\phi(z) = 0$. The solution to this equation is optimally computed using the fast marching algorithm [8]. Once ϕ is computed, the computation of the geodesic can be reformulated as a backtracking procedure following the gradient of ϕ

$$\Gamma_{\psi,y} = \{x \in S | \nabla\Gamma_{\psi,y}(x) = \nabla_S\phi(x), \Gamma_{\psi,y}(0) = y, \Gamma_{\psi,y}(1) = z\}, \qquad (1)$$

where $\nabla_S\phi(x)$ denotes the gradient of ϕ intrinsic to the surface. This means that starting from any point y, the geodesic on the surface induced by ψ is obtained following the gradient of ϕ. Two considerations have to be taken into account. First, the surface S has to be a good approximation of a Riemmanian manifold, in other words, be smooth enough to allow computing its gradient. Secondly, $\nabla_S\phi(x)$ has to be defined for all x on S. Singular points, such as sink or source points are excluded from the path computation, however, they will provide important information as we will show later. Notice that if the weighted map ψ is constant, we are in the Euclidean case. However, it is interesting for other applications the use of other maps defined on the surface. In our case, ϕ is directly given by the LAT map, where z is located at the HIS, and therefore, direct application of equation 1 (i.e. backtracking) will provide the geodesics. Notice that the geodesics will be calculated from late local activation times to early activation times, thus using the negative of the gradient field, $-\nabla_S\phi$. Our implementation will follow the description given in [7], where the computations are performed on implicit surfaces instead of on triangular meshes.

2.2   End-Terminals Detection

The end-terminals are estimated directly from the electro-anatomical maps. Observing closely the gradient field of a simulated electro-anatomical map (Fig. 1 right) one can clearly distinguish points where the geodesics converge or diverge, which are singular points of the map, i.e. points where the gradient field of the map is not well defined. Therefore, a point x on a distance map ϕ defined on the surface S will be detected as a singular point if the lateral derivatives of the gradient map are different at every tangential direction. In practice, the singular points are detected when the sum of these derivatives are below a certain threshold:

$$\lim_{h \to 0^+} \frac{\nabla_{u_i}\phi(x) - \nabla_{u_i}\phi(x - u_i h)}{h} + \lim_{h \to 0^-} \frac{\nabla_{u_i}\phi(x) - \nabla_{u_i}\phi(x - u_i h)}{h} \lessgtr \varepsilon,$$

where u_i are the coordinates of the tangential plane of S defined on x, and $\nabla_{u_i}\phi$ the gradient component in this direction. The threshold selection will be studied in Section 3.2.

A distinction has to be made between two types of singular points, those where geodesics converge, denoted as sink points (point A in Fig. 1) and source points where geodesics diverge (point B in Fig. 1). These points can be discriminated based on the change of sign of $\nabla_{u_i}\phi$ along each tangential direction. A change from positive to negative gives sink points and source points otherwise. Our points of interest here are the sink points of the activation map $\phi(x)$, that will be equivalent to source points of the negative map $-\phi(x)$.

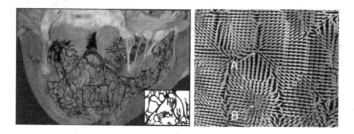

Fig. 1. Left: detail of the Purkinje fiber system in a real case. Right: vector field of an electrical activation map. Two singular points are indicated with arrows, a sink point: A, and a source point: B.

3 Validation Experiments with Simulated Data

3.1 Fast Marching Generated Simplistic Map

To show the behavior of the proposed technique we have designed a simple simulation experiment with synthetic data. Notice that the simulated map is not meant to be realistic but it is designed to show the tracking method performance. We have taken a surface model of a left ventricle, S, and a tree mimicking its main PK system defined on it. Starting from a tree point located at the HIS, a signal is propagated to the rest of points in the PK tree with a constant velocity v_1. Starting from the local activation times obtained on this tree, a signal is propagated using fast marching to the rest of the surface points with a lower velocity, $v_2 = v_1/10$, obtaining a simulated activation map $\phi(\boldsymbol{x}) \ \forall \ \boldsymbol{x} \in \mathcal{S}$. Fig. 2 (left) shows the reference tree with the surface mapped with the local activation times obtained in this way. Our goal in this experiment is to use this map to recover the initial tree with the backtracking algorithm described above. In Fig. 2 (middle), we show the geodesic paths obtained starting from a set of seeds equally distributed on the surface (2478 seeds in total). This gives us information about many possible activation paths on the ventricle. Then, calculating the velocity along the generated paths and keeping the ones with high velocities we are able to recover the original PK tree as shown in Fig. 2 (right). We have then computed the LAT map from the estimated tree to compare it with the original LAT map. This map is shown in figure Fig. 2 (middle) and the absolute differences between both are shown in Fig. 2 (right), with a maximum difference of 6.4 ms and an average relative error of 0.4 %. This figure shows that the bigger differences are obtained close to the estimated tree, particularly at gaps.

Fig. 2. Ventricle surface with simulated activation maps. Left: a reference (simulated) PK tree is shown from which the activation map is computed. Middle: geodesic paths estimated from the simulated map using uniformly distributed seeds over the surface (shown in red), and map generated from the filtered tree. Right: differences between initial LAT and LAT computed with the estimated filtered tree, shown in white.

3.2   Realistic Simulated Data

A detailed electrophysiolocal model including a complete Purkinje tree has been used with the method described in [2], resulting in simulated electrical maps of the left ventricle. The method is an enhancement of a rule-based method known as the Lindenmayer systems (L-systems). The construction of the PK tree is divided into three consecutive stages, which subsequently develop the CCS from proximal to distal sections. Each stage is governed by a set of independent user parameters together with anatomical and physiological constraints to direct the generation process and adhere to the structural observations derived from histology studies. Several properties of the tree are defined using statistical distributions to introduce stochastic variability in the models. The CCS built with this approach can generate electrical activation sequences with physiological characteristics. The electrical propagation in the myocardium was modeled using the monodomain equation.

Using these maps, we have tested our end-terminals detection, by finding the singular points, depicted in Fig. 3 (middle), and comparing them with the true end-terminals used in the simulations, shown in Fig. 3 (left). For quantitative evaluation, the distances from the detected points and the reference points are computed. To account for under or overestimation of the detected points with respect to the threshold used, two distance measures are computed: the average distance computed from each reference point to its nearest detected point; and viceversa, the average distance computed from each detected point to its nearest reference point. The optimal threshold value is taken when these two distances reach low values at the same time. From Fig. 3 (right) the optimal threshold is found at approximately 0.35, providing average distances of 0.98 and 0.88 mm respectively.

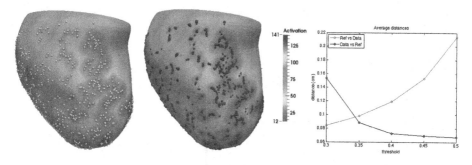

Fig. 3. Left: synthetically generated electrical map using [2], with end-terminals shown in white. Middle: automatically detected end-terminals shown in blue for $\varepsilon = 0.35$. Left: Average differences between the reference end-terminals and the detected singular points and viceversa vs the threshold value, ε.

4   Clinical Data Results

We had access to clinical data for one case of fascicular VT, where the Cardiac Conduction System is thought to play a relevant role. Specifically a pre-operative CT image was available and used to extract the LV geometry and electro-anatomical mapping data (CARTO, Biosense Webster) giving intra-cardiac electrical information at the LV endocardium. The mapping of the CARTO data onto the CT geometry was based on establishing an homeomorphism between both surfaces using a common parameterization computed by mesh flattening (see details in [9]). The electro-anatomical maps is composed of 231 points where a 1D electrocardiogram signal (acquired at 1 kHz) is available for each point during approximately three cardiac cycles. Purkinje and muscle activations were visually identified by an experienced technician in Hospital Clínic de Barcelona on these 1D signals. The final reference surface mesh based on the CT geometry has 50k nodes and LAT values from CARTO are linearly interpolated.

Fig. 4 shows the geodesic paths automatically estimated for this real case, generated from the singular sink points detected by our method from the negative gradient field of the activation map, $-\nabla_S\phi$. The LV is colored with the activation map in two different views (A, B). The geodesic paths have also been colored according to the velocity, see Fig. 4 (E, F), in the same views for comparison. Notice that a discrimination between fast or slow conduction channels is difficult or impossible in the clinical case because the conduction velocity considerably changes along every path. However, it is interesting to see regions with consistently higher velocity, giving a hint about possible contacts with the PK system.

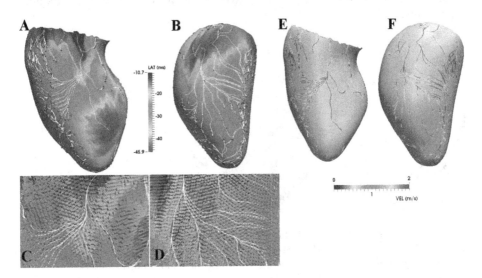

Fig. 4. A, B: LV surface from a real patient with overlaid measured activation maps, computed geodesic paths (white), and seeds (red). C, D: detail of the geodesic paths with the gradient vector field obtained from the activation maps. E, F: geodesics colored according to their conduction velocity.

5   Conclusions

We have proposed a method to automatically trace the electrical pathways in activation maps of the LV surface. It is important to remind that these electrical pathways are just the pathways of electrical activation through the muscle, and are not meant to be an exact reconstruction of the PK system. Taking this into account, the geodesic paths obtained here have several advantages over the standard electrical maps. First, they provide an attractive and an alternative visualization to the activation maps, giving more detailed local information. For instance, we can better see channels forming loops, or branching systems, that are sometimes hidden by the global color-coded maps, and that are useful to the clinicians to understand and treat abnormal electrical patterns. Second, the velocity information along the geodesic paths give us an approximate idea about how is the distribution of the PK system, providing information about the connections of the conduction paths of the muscle to the PK system. The geodesics shown here open up the possibility for better visualizations in other cases of VT and can be used as an alternative way to identify scar tissues.

Acknowledgements. This work is partially funded by the Sub-programa de Proyectos de Investigación en Salud Instituto de Salud Carlos III, Spain (FIS - PI11/01709), by Spanish Ministry of Science and Innovation (TIN2011-28067), and by eTorso project (2013-001404) from Generalitat de Valencia.

References

1. Piers, S., van Huls van Taxis, C., Tao, Q., van der Geest, R., Askar, S., Siebelink, H.M., Schalij, M., Zeppenfeld, K.: Epicardial substrate mapping for VT ablation in patients with nonischaemic cardiomyopathy: a new algorithm to differentiate between scar and viable myocardium developed by simultaneous integration of CT and contrast-enhanced MR imaging. European Heart J. 34, 586–596 (2013)
2. Sebastian, R., Zimmerman, V., Romero, D., Sanchez-Quintana, D., Frangi, A.: Characterization and modeling of the peripheral cardiac conduction system. IEEE Trans. Med. Imaging 32(1), 45–55 (2013)
3. Durrer, D., van Dam, R.T., Freud, G.E., Janse, M.J., Meijler, F.L., Arzbaecher, R.C.: Total excitation of the isolated human heart. Circulation 41(6), 899–912 (1970)
4. Dijkstra, E.: A note on two problems in connection with graphs. Numerische Math. 1, 269–271 (1959)
5. Kimmel, R., Sethian, J.A.: Computing geodesic paths on manifolds. Proc. Nat. Acad. Sci. 95(15), 8431–8435 (1998)
6. Martínez, D., Velho, L., Carvalho, P.: Computing geodesics on triangular meshes. Computers & Graphics 29(5), 667–675 (2005)
7. Mémoli, F., Sapiro, G.: Fast computation of weighted distance functions and geodesics on implicit hyper-surfaces. Journal of Computer Physics 173, 730–764 (2001)
8. Sethian, J.A.: A fast marching level-set method for monotonically advancing fronts. Proc. Nat. Acad. Sci. 93, 1591–1595 (1996)
9. Soto-Iglesias, D., Butakoff, C., Andreu, D., Fernández-Armenta, J., Berruezo, A., Camara, O.: Evaluation of different mapping techniques for the integration of electro-anatomical voltage and imaging data of the left ventricle. In: Ourselin, S., Rueckert, D., Smith, N. (eds.) FIMH 2013. LNCS, vol. 7945, pp. 391–399. Springer, Heidelberg (2013)

Velocity-Based Cardiac Contractility Personalization with Derivative-Free Optimization

Ken C.L. Wong[1], Maxime Sermesant[2,3], Jatin Relan[2], Kawal S. Rhode[3],
Matthew Ginks[3], C. Aldo Rinaldi[3], Reza Razavi[3], Hervé Delingette[2],
and Nicholas Ayache[2]

[1] Computational Biomedicine Laboratory, Rochester Institute of Technology, Rochester, USA
kenclwong@gmail.com
[2] INRIA, Asclepios Project, 2004 Route des Lucioles, Sophia Antipolis, France
[3] King's College London, St Thomas' Hospital, Division of Imaging Sciences, London, UK

Abstract. Cardiac contractility personalization from medical images is a major step for biophysical models to impact clinical practice. Existing gradient-based optimization approaches show promising results of identifying the maximum contractility from images, but the contraction and relaxation rates are not accounted for. A main reason is the limited choice of objective functions when their gradients are required. For complicated cardiac models, analytical evaluation of the gradient is very difficult if not impossible, and finite difference approximation may introduce numerical difficulties and is computationally expensive. We remove such limits by using derivative-free optimization, and propose a velocity-based objective function on identifying the maximum contraction, contraction rate, and relaxation rate simultaneously with intact model complexity. Experiments on synthetic data show that the parameters are better identified using the velocity-based optimization than the position-based one. Experiments on clinical data show that the framework can obtain personalized contractility consistent to the physiologies of the patients.

1 Introduction

Cardiac model personalization is a process to obtain a biophysical model accounting for the subject-specific cardiac physiology, usually realized as parameter estimation. Given a generic cardiac model designed from invasive experiments, model parameters of anatomy, electrophysiology or mechanics are estimated from the subject-specific *in vivo* measurements such as non-contact endocardial mappings and magnetic resonance images (MRI). As simulation of the whole organ has reached a degree of realism which is quantitatively comparable with available cardiac images and signals acquired routinely on patients, model personalization gives a potential impact to clinical practice by improving disease diagnoses and planning therapies.

Cardiac mechanics is the interaction among active contraction, tissue stiffness, and boundary conditions of surrounding anatomical structures [1]. Various cardiac electromechanical models have been proposed to describe such an interaction with different physiological plausibilities, complexities, and computational efficiencies. According to the characteristics of the models, different personalization algorithms have been proposed to estimate tissue stiffness and active contraction properties [2–4].

O. Camara et al. (Eds.): STACOM 2013, LNCS 8330, pp. 228–235, 2014.

This paper will concentrate on cardiac contractility personalization. To estimate model parameters, the gradient-based optimization is probably the mostly used method for the variational approach [5–7]. For example, in [7], the quasi-Newton L-BFGS-B algorithm was utilized to optimize the position-based objective function, which requires the gradient of the objective function with respect to the contraction parameters. Although the utilized adjoint method allows efficient computation of the gradient, it requires the system derivatives of the complicated cardiac electromechanical model. This limits the exploration of the proper objective functions and also the types of parameters to be estimated, as some objective functions are highly nonlinear with respect to the desired parameters. Therefore, only the maximum contraction was estimated even after some model simplifications.

In consequence, we propose the use of derivative-free optimization for cardiac contractility personalization. Without the analytical, numerical, and computational difficulties associated with gradient evaluation, objective functions which may provide better parameter estimation can be investigated with relative ease. By using the derivative-free optimization method based on trust region methods [8], we propose a velocity-based objective function for simultaneous estimation of regional maximum contraction, rate of contraction, and rate of relaxation. Experiments were performed on synthetic data to show the capability of the framework in identifying regional parameters, and on patient data to show the clinical relevance.

2 Cardiac Electromechanical Model

The dynamics of a cardiac electromechanical model can be given as:

$$\mathbf{M}\ddot{\mathbf{U}} + \mathbf{C}\dot{\mathbf{U}} + \mathbf{K}\mathbf{U} = \mathbf{F}_b + \mathbf{F}_c \tag{1}$$

where $\mathbf{M}$, $\mathbf{C}$, and $\mathbf{K}$ are the mass, damping, and stiffness matrices, and $\ddot{\mathbf{U}}$, $\dot{\mathbf{U}}$, and $\mathbf{U}$ are the acceleration, velocity, and displacement vectors. $\mathbf{F}_b$ is the external load vector of boundary conditions, comprising the simulated blood pressures and the displacement constraints. $\mathbf{F}_c$ is the active contraction force vector derived from electrophysiology and the tissue structure. The electromechanical model in [9] is used in this paper.

To obtain $\mathbf{F}_c$ at any point of the myocardium, the relation between the action potential and the active contraction can be modeled as [9]:

$$\begin{cases} \sigma_c(t) = \sigma_0(1 - e^{\alpha_c(T_d-t)}) & \text{if } T_d \leq t \leq T_r \\ \sigma_c(t) = \sigma_c(T_r)e^{\alpha_r(T_r-t)} & \text{if } T_r < t < T_d + HP \end{cases} \tag{2}$$

with σ_c the contraction stress and HP the heart period. σ_0 is the maximum contraction, and α_c and α_r are the contraction and relaxation rates to control the change of σ_c. T_d and T_r are the depolarization and repolarization times derived from the action potential, and a time constant can be added to model the delay between the electrical and mechanical phenomena. Therefore, the parameters of interest are σ_0, α_c, and α_r.

3 Electrophysiology and Kinematics Personalization

To avoid accumulating sources of uncertainties, patient-specific datasets including a rich description of cardiac electrophysiology were utilized. In addition to the acquisition of anatomical and cine MRI, non-contact endocardial mappings have been acquired. The extracted depolarization and repolarization isochrones then serve as input information to an electrophysiology personalization method [10] which minimizes the discrepancy between measured and simulated isochrones, providing the electrical propagation for both kinematics and mechanics personalization.

Kinematics personalization consists in estimating the motion of cardiac structures from images. The kinematics personalization approach in [5] is used with the cardiac electromechanical model in Section 2. The two ventricles are meshed with tetrahedra from the anatomical MRI (Fig. 1). The evolution of the displacement of each mesh node is governed by (1) with an embedded image force:

$$\mathbf{M}\ddot{\mathbf{U}} + \mathbf{C}\dot{\mathbf{U}} + \mathbf{K}\mathbf{U} = \mathbf{F}_b + \mathbf{F}_c + \beta\mathbf{F}_{\text{img}} \tag{3}$$

In the kinematics personalization, $\mathbf{F}_{\text{img}}$ corresponds to a force vector which tracks salient image features in the image sequence, computed using a 3D block-matching algorithm to attract points towards the nearest edge voxels. Image forces are not physiology based since their sole purpose is to help tracking the cardiac motion, and they are discarded during the mechanical personalization. The personalized nodal positions and velocities are obtained from the kinematics personalization, which are used as the inputs for contractility personalization, along with the personalized electrophysiology.

4   Mechanics Personalization with Derivative-Free Optimization

Kinematics personalization produces cardiac motion consistent with the apparent motion in the images. Nevertheless, it cannot address the underlying physiological properties of the patient, such as the active contraction properties. To infer the physiological properties, mechanics personalization is required. To reduce the complexity of the problem, we only concentrate on the active parameters.

4.1   Objective Function

The similarity between simulations and measurements is defined by an objective function. Supposing that the heart geometry is partitioned into regions, the objective function for variational data assimilation can be given as:

$$\mathcal{F}(\boldsymbol{\theta}) = \sum_k \sum_r \left(\frac{\sum_i \|\bar{\mathbf{y}}_{k,i} - \mathbf{y}_{k,i}(\boldsymbol{\theta})\|^2}{n_r} \right) \tag{4}$$

where $\boldsymbol{\theta}$ is a vector comprising parameters $(\sigma_0, \alpha_c, \alpha_r)$ of all regions. $\bar{\mathbf{y}}_{k,i}$ is the measurement at discrete time instant k of point i in region r, and $\mathbf{y}_{k,i}(\boldsymbol{\theta})$ is the corresponding simulated quantity. n_r is the number of measurements in a region, which can be used to remove the bias towards regions with more measurements.

Different types of measurements $\bar{\mathbf{y}}_{k,r}$ can give different results. In [5, 7], the positions of the personalized cardiac kinematics were used. As only the maximum contraction parameters σ_0 were estimated, the sole use of measured positions may provide meaningful results. Nevertheless, if α_c and α_r are also desired, positions alone may

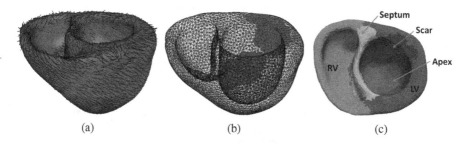

Fig. 1. Heart representation. (a) Heart geometry and fiber orientations. (b) Scar regions. (c) 5-region representation.

not provide the necessary temporal information. Furthermore, it has been shown using control theory that velocity-based data assimilation can lead to a more stable system [11]. Therefore, we use velocities instead.

4.2 Bound Constrained Optimization without Derivatives

To estimate the contraction parameters, the optimization problems were solved using gradient-based algorithms on synthetic data in [5, 6] and on clinical data in [7]. Analytical computation of the gradient requires the derivatives of the electromechanical model. These derivatives are difficult to derive as the model involves interactions between myocardial deformation, contraction stresses, and different boundary conditions. Therefore, it is difficult to compute the gradient analytically without making significant simplifications which sacrifice the model integrity and thus the estimation accuracy. On the other hand, finite difference is a popular numerical alternative when analytical evaluation of the gradient is infeasible. Nevertheless, the associated computational complexity is impractical to our problem, and may also introduce further numerical difficulties and instability.

In view of these issues, the BOBYQA algorithm for derivative-free optimization is utilized in this framework [8]. The basic idea is to approximate the curvature of the objective function by forming a quadratic model using interpolation. Let n be the number of parameters to be estimated. By providing the initial parameters, and also the beginning and ending trust regions, the algorithm forms a quadratic model Q by computing the values $\mathcal{F}(\boldsymbol{\theta}_i)$ of $2n+1$ interpolation points $\boldsymbol{\theta}_i$ within the beginning trust region. With the quadratic model available, its minimum point $\bar{\boldsymbol{\theta}}$ can be determined. $\mathcal{F}(\bar{\boldsymbol{\theta}})$ is then computed to verify if Q is a good local approximation of $\mathcal{F}$, and Q and the trust region are updated using the information. These procedures iterate until the trust region is smaller than the ending trust region, which defines the desired preciseness of the estimation. Therefore, explicit computation of the gradient is not required in BOBYQA. This gives larger flexibility in choosing objective functions and parameters, as we do not need to consider how to obtain the system derivatives. Moreover, as each iteration only requires one function evaluation of $\mathcal{F}(\bar{\boldsymbol{\theta}})$, the computational load is much lower than that of the finite difference, with requires at least n function evaluations.

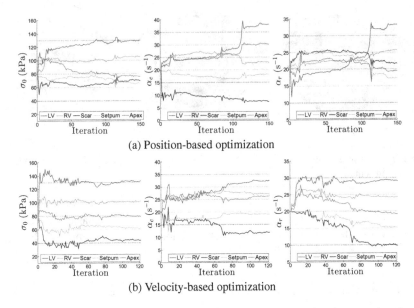

(a) Position-based optimization

(b) Velocity-based optimization

Fig. 2. Synthetic data. Estimated parameters, with dotted lines representing the ground truth values. Left to right: maximum contraction σ_0, contraction rate α_c, and relaxation rate α_r.

5   Experiments

5.1   Evaluation on Synthetic Data

Experimental Setups. The heart representation was created from the data of a patient with myocardial infarction. The heart geometry was segmented from the image frame at mid-diastole, and a FEM mesh with synthetic fiber orientations was obtained with known infarcted regions identified by clinicians through late-enhancement MRI (Fig. 1(a) and (b)). The personalized T_d and T_r derived from the patient noncontact endocardial electrical mappings were used in (2) [10]. The simulated positions and velocities on the heart surfaces were used as the inputs to the experiments, with initial parameters $\sigma_0 = 80$ kPa, $\alpha_c = \alpha_r = 20$ s^{-1}.

Results. Fig. 2 shows the estimation results. For the position-based optimization, most estimated parameters are inaccurate, and the regional orders cannot reflect those of the ground truth. On the other hand, the velocity-based optimization has correct regional orders and more accurate results.

Fig. 3 shows the mean position differences between the ground truth and the personalized simulations in a cardiac cycle. Although all differences are below the usual spatial resolution of MRI (< 0.6 mm), the velocity-based optimization has smaller differences and variations. These results show that the velocity-based optimization performs better than the position-based one.

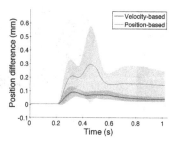

Fig. 3. Synthetic data. Mean position differences between the ground truth and the personalized simulations in a cardiac cycle. The shaded area represents the standard deviation.

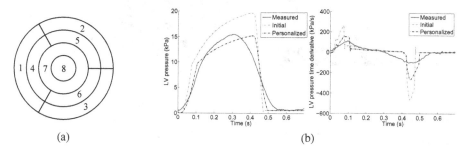

(a)   (b)

Fig. 4. Patient data. (a) Patient 2. LV regions based on AHA nomenclature. (b) Patient 1. Results of velocity-based optimization. Left: LV pressures. Right: the corresponding time derivatives.

5.2   Evaluation on Clinical Data

Experimental Setups. Two data sets were tested. Patient 1 has myocardial infarction, whose data set was used in Section 5.1 to generate the synthetic data, with infarcted regions identified through late-enhanced MRI. Patient 2 has dilated myocardiopathy without identified infarction. All patients have left bundle branch block (LBBB) and suffer from heart failure. Each data set has a cine MRI sequence of 30 frames, with the heart periods of Patient 1 and 2 as 1.03 and 0.73 s respectively. The corresponding in-plane resolutions are 1.56 and 1.45 mm^2, and all images have inter-slice resolution of 10 mm. All data sets have the endocardial activation maps measured with the Ensite balloon, which were extrapolated to the myocardial volume using an electrophysiological model to provide the subject-specific T_d and T_r in (2) for the experiments [10].

For each data set, the heart geometry was segmented from the image frame at mid-diastole, and a FEM mesh with synthetic fiber orientations was constructed. For Patient 1, the 5-regional heart representation in Section 5.1 was used (Fig. 1). For Patient 2, in our early experiments, the heart geometries were only divided into LV and RV. Nevertheless, the personalized simulations were inconsistent with the measurements because of the local variations possibly caused by the diseases. Therefore, the LV was divided into eight regions by grouping the American Heart Association (AHA) regions to balance between preciseness and computational load (Fig. 4(a)). Kinematics

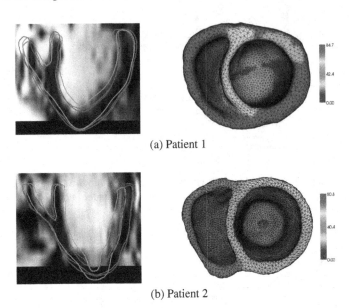

(a) Patient 1

(b) Patient 2

Fig. 5. Patient data at end systole. Left: heart geometries overlapped with images, with red, green, and blue representing personalized kinematics, initializations, and personalized simulations respectively. Right: personalized simulations and the corresponding contraction stresses in kPa.

Table 1. Patient data. Estimated contraction parameters.

Patient 1						Patient 2								
Region	LV	RV	Scar	Sep	Apex	Region	1	2	3	4	5	6	7	8	RV
σ_0 (kPa)	84.7	85.0	54.2	60.0	49.3	σ_0 (kPa)	52.3	56.6	55.5	97.7	54.6	62.9	35.9	122.0	83.1
α_c (s^{-1})	27.2	11.8	21.3	22.2	20.8	α_c (s^{-1})	23.8	29.0	11.0	25.8	33.7	25.9	30.5	20.2	5.8
α_r (s^{-1})	33.2	26.4	30.1	7.9	26.5	α_r (s^{-1})	31.8	40.9	25.7	27.1	39.5	26.9	36.3	14.0	14.4

personalization was performed on each data set to provide positions and velocities for the experiments. Only the points on the heart surfaces were used as motion information is unavailable inside the myocardium for cine MRI. The initial contraction parameters were $\sigma_0 = 100$ kPa, $\alpha_c = \alpha_r = 30$ s^{-1}.

Results. As we have shown that the velocity-based optimization is better on the synthetic data, it will be the concentration in the following discussions.

Fig. 5 provides the comparisons among the personalized kinematics, simulations with initial parameters, and personalized simulations. In all cases, the simulations with the personalized parameters are much closer to the personalized kinematics. Fig. 5 also shows the active contraction stresses at the end of systole, which correspond to the maximal contractility. In all cases, the RV has larger contractility compared with most regions at the LV. For Patient 1, the scar region has lower contractility, which is consistent to the pathology. Nevertheless, the septum also has lower contractility even it is not infarcted. In fact, septal flash can be observed in the image sequence, thus the low contractility partially accounts for the corresponding condition. For the apex, as a large part of this region is out of the image, it is hard to justify the clinical relevance of its contractility. For Patient 2, the small contractility of the LV adequately reflect the symptom of dilated cardiomyopathy. The inconsistently high contractility of the apex was mainly caused by the imposed displacement boundary conditions, and its location outside of the image region. The estimated parameters are shown in Table 1.

To show the realism of the personalized mechanics, the simulated blood pressures through ventricular isovolumetric constraints and Windkessel model are compared with the invasively measured blood pressure of the patients (Fig. 4(b)). The LV blood pressure and its time derivative show large improvement after mechanics personalization. This means that the personalized electromechanical model can partially reflect the subject's actual physiology.

References

1. Germann, W.J., Stanfield, C.L.: Principles of Human Physiology. Pearson Benjamin Cummings (2005)
2. Hu, Z., Metaxas, D., Axel, L.: In vivo strain and stress estimation of the heart left and right ventricles from MRI images. Medical Image Analysis 7(4), 435–444 (2003)
3. Wang, V.Y., Lam, H.I., Ennis, D.B., Cowan, B.R., Young, A.A., Nash, M.P.: Modelling passive diastolic mechanics with quantitative MRI of cardiac structure and function. Medical Image Analysis 13(5), 773–784 (2009)
4. Xi, J., Lamata, P., Lee, J., Moireau, P., Chapelle, D., Smith, N.: Myocardial transversely isotropic material parameter estimation from in-silico measurements based on reduced-order unscented Kalman filter. Journal of the Mechanical Behavior of Biomedical Materials 4(7), 1090–1102 (2011)
5. Sermesant, M., Moireau, P., Camara, O., Sainte-Marie, J., Andriantsimiavona, R., Cimrman, R., Hill, D.L.G., Chapelle, D., Razavi, R.: Cardiac function estimation from MRI using a heart model and data assimilation: advances and difficulties. Medical Image Analysis 10, 642–656 (2006)
6. Sundar, H., Davatzikos, C., Biros, G.: Biomechanically-constrained 4D estimation of myocardial motion. In: Yang, G.-Z., Hawkes, D., Rueckert, D., Noble, A., Taylor, C. (eds.) MICCAI 2009, Part II. LNCS, vol. 5762, pp. 257–265. Springer, Heidelberg (2009)
7. Delingette, H., Billet, F., Wong, K.C.L., Sermesant, M., Rhode, K., Ginks, M., Rinaldi, C.A., Razavi, R., Ayache, N.: Personalization of cardiac motion and contractility from images using variational data assimilation. IEEE Transactions on Biomedical Engineering 59(1), 20–24 (2012)
8. Powell, M.J.D.: The BOBYQA algorithm for bound constrained optimization without derivatives. Technical report, DAMTP, University of Cambridge (2009)
9. Sermesant, M., Delingette, H., Ayache, N.: An electromechanical model of the heart for image analysis and simulation. IEEE Transactions on Medical Imaging 25(5), 612–625 (2006)
10. Relan, J., Chinchapatnam, P., Sermesant, M., Rhode, K., Ginks, M., Delingette, H., Rinaldi, C.A., Razavi, R., Ayache, N.: Coupled personalization of cardiac electrophysiology models for prediction of ischaemic ventricular tachycardia. Journal of the Royal Society Interface Focus 1(3), 396–407 (2011)
11. Moireau, P., Chapelle, D., Le Tallec, P.: Joint state and parameter estimation for distributed mechanical systems. Computer Methods in Applied Mechanics and Engineering 197(6-8), 659–677 (2008)

Model-Based Estimation of 4D Relative Pressure Map from 4D Flow MR Images

Viorel Mihalef[1], Saikiran Rapaka[1], Mehmet Gulsun[1], Angelo Scorza[1],
Puneet Sharma[1], Lucian Itu[1], Ali Kamen[1], Alex Barker[2], Michael Markl[2],
and Dorin Comaniciu[1]

[1] Imaging and Computer Vision, Siemens Corporation, Corporate Technology,
Princeton, New Jersey
[2] Departments of Radiology and Biomedical Engineering, Northwestern University,
Feinberg School of Medicine, 737 N. Michigan Avenue Suite 1600 Chicago, Illinois

Abstract. We propose a new framework for 4D relative pressure map
computations from 4D flow MRI that uses enhanced geometric models
for the blood vessels and flow-aware surface and volumetric tags. The
enhanced geometric modeling provides better accuracy compared to a
simple voxelized mask, while tagging of inlets and outlets allows imposing physiologically meaningful boundary conditions, contributing to
more accurate pressure computations. An integrated software suite for
semi-automatic processing of 4D flow MR images, preparation and computation of the flow parameters is presented. This enables a fast and
intuitive workflow, with accurate final results, ready in minutes.

1 Introduction

Knowledge of pressure and velocity of blood flow in the human cardiovascular
system can be decisive for clinical evaluations (initial and post-procedural) and
procedure planning. In particular, the severity of various cardiovascular diseases
such as aortic valve stenosis and aortic coarctation can be assessed from intraluminal pressure gradients [3]. Non-invasive time-resolved 3D phase contrast
(PC) MRI with three-directional velocity encoding (also termed as 4D flow MRI)
[11,12] provides in-vivo blood velocity information that can be used to derive
intraluminal relative pressures. This approach offers a full 3D+time computation of the relative pressure, which can be used to estimate temporal and spatial
gradients within a vessel segment. In this work we present an efficient workflow
for the computation of relative pressure from 4D flow MRI.

Previous works [1,15,6] that consider the issues of velocity reconstruction and
pressure map estimation from PC-MRI data, use, in a first step, various techniques which essentially act as enhancement filters, in order to enforce incom-
pressibility of the given velocity field. Such filters are either global [14] or, more
commonly, have compact support [2,13,10,5] given by voxel masks approximat-
ing the region of interest. In a second step, the pressure map is computed from
the filtered velocity field by solving the Pressure Poisson Equation (PPE) with
either Neumann or Dirichlet boundary conditions. The numerical formulations

O. Camara et al. (Eds.): STACOM 2013, LNCS 8330, pp. 236–243, 2014.

vary between Cartesian methods (finite differences or finite volumes) and finite element methods (FEM). Another class of methods compute the time-varying pressure drop by computing Navier-Stokes flow in the vessel, using aortic geometries modeled either as 3D rigid walls, or as axisymmetric 1D full FSI model with Windkessel boundary conditions [9].

In this paper, we propose a method that addresses issues with some of the simplifying assumptions used in the previous work, that may introduce errors in the estimation of relative pressure. For example, we go beyond models which use streamline/pathline pressure integrals, which offer information insufficient for determining cross-sectional gradients. We also note that, if the boundary conditions do not recognize the different types of boundaries that the computational cells cross, wall fluxes could be easily overestimated or underestimated, and small changes in the lumen mask may introduce significant flux changes. Krittian et. al. ([10]) recently proposed a method that mitigates such issues by rewriting the PPE in weak form and solving a volume integral equation using FEM. However, flux mass conservation may not be enforced properly if the specific branch fluxes are not taken into account. Furthermore, using the same mask for all time phases may shrink the domain, making it harder to enforce mass conservation.

Our approach for filtering the initial velocity field and solving the PPE with Neumann boundary conditions improves on the state of the art by using enhanced geometric models for the blood vessels that are an order of accuracy above the voxelized mask, and include flow-aware tagging of inlet and outlet regions, which allows flux constraints. We built an integrated framework for data processing and computation that enables the user to progress through the data processing pipeline in a fast, semi-automatic, intuitive fashion, and to obtain the final result in a matter of minutes. Our pressure computation module is refined but efficient at the same time, providing the complete 4D results in less than a minute for a usual 4D PC-MRI dataset.

2   Relative Pressure from Velocity Fields: Theory

Blood in larger vessels can be appropriately modeled as an incompressible Newtonian fluid, whose flow inside a given moving domain can be modeled by the classical Navier-Stokes equations [7]

$$\rho\left(\mathbf{u}_t + (\mathbf{u} - \mathbf{w}) \cdot \nabla \mathbf{u}\right) = -\nabla p + \mu \Delta \mathbf{u} + \rho \mathbf{F}, \tag{1}$$
$$\nabla \cdot \mathbf{u} = 0 \tag{2}$$

where t is time, ρ is the fluid density, $\mathbf{u}$ is the fluid velocity, $\mathbf{w}$ is the reference domain velocity, p is the fluid pressure, μ is the dynamic viscosity, and $\mathbf{F}$ is the sum of external forces such as gravity. By taking the divergence of equation 1, one obtains the Pressure Poisson Equations (PPE). Further, projecting equation 1 along the normal to the domain boundary, we obtain the natural (Neumann) boundary condition:

$$\Delta p = \nabla \cdot RHS, \quad \frac{\partial p}{\partial n} = RHS \cdot \hat{n} \tag{3}$$

where $RHS = -\rho\left(\mathbf{u}_t + (\mathbf{u} - \mathbf{w}) \cdot \nabla \mathbf{u}\right) + \mu \Delta \mathbf{u} + \rho \mathbf{F}$. In this work we ignore the role of gravity, due to horizontal patient positioning, which mitigates its effect on measurements. While the Neumann boundary condition is the natural choice when the velocity is incompressible, this is not necessarily the case with measured data, and Dirichlet boundary conditions for pressure can be used on vessel outlets whose fluxes are known.

3   Relative Pressure from Velocity Fields: Methodology

As the quality of 4D flow MRI improves, one can compute progressively better geometric models of vessel geometries, and luminal velocity with increased accuracy. In particular, recently [8] have presented, as part of an investigational 4D Flow tool (Siemens AG), a methodology for extracting a 4D geometric model of the aorta and deformable masks from 4D flow MRI thoracic data, including the associated 4D aortic flow information. We used the 4D Flow tool as a baseline for our workflow, and added further functionality to compute tagging information for the aortic walls, virtual inlets and outlet locations, as well as a level set representation for the mask. Our computational domain is therefore computed with enhanced accuracy, which further allows more accurate application of the boundary conditions for the pressure Poisson solver. In particular, the level set allows computation of enhanced edge weights for the Poisson solver.

Our approach is summarized in the schematic and flowchart given in Figure 1. In the following we discuss in more detail each of these components.

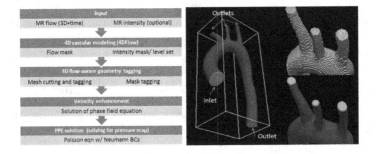

Fig. 1. Left: flowchart depicting the workflow from image acquisition to relative pressure computation. Right: our geometric setup: (left image) wall surface defined as the zero of a level set; inlet/outlet tagging included. Previous methods use voxelized data (right top) while we use smooth tagged representations (right bottom).

4D Intensity and Velocity Masks. 4D flow data are imported into the 4D Flow Tool, and a vascular model at a reference time point is obtained by performing aortic lumen segmentation following a semi-automated centerline extraction [8]. This vascular model is propagated across the entire time sequence using the displacement fields derived from a deformable registration technique.

Flow-Aware Geometric Tagging. While standard finite difference spatial discretization of the PPE computational domain relies on low order accuracy voxelized masks, with staircase boundary, our computational mask is defined implicitly by a level set, and therefore provides a smooth wall surface. Furthermore, we introduce a "flow-aware" semi-automatic procedure to tag the inlets and outlets of the vessel mesh, as well as the inlet and outlet cells and nodes of the computational domain (Figure 1). The semi-automatic workflow proceeds as follows: the user clicks on the desired position where the virtual cut is to be placed, and prescribes the appropriate tag corresponding to an inlet or an outlet. The virtual cut normal plane is computed using connected component separation of all the mesh triangles intersected by the plane, followed by an area cut minimization procedure. The non-inlet/outlet triangles are tagged as walls. A level set function is subsequently computed as the signed-distance to the mesh. In the next step, wall, inlet and outlet cells are tagged (based on their intersection with the tagged mesh), followed by nodal tagging based on the cell tagging and level set values. The inlet/outlet cuts are then propagated both as topological and geometric cuts in time, using the displacement fields from section 3.1, and Taubin-like curve-smoothing in order to minimize distortion and preserve flatness. The outcome is a set of 4D tagged meshes with point-correspondence, and corresponding 4D grids with tagged cells and nodes (solid/fluid/inlet/outlet) and level set information.

Velocity Field Enhancement Prior to PPE Solution. The velocity field measured in 4D flow MRI is not discretely incompressible, and suffers from various aliasing and noise artifacts, especially near the wall, and various methods have been proposed to address such issues. In our framework, we perform a correction step prior to solving the PPE by setting up a Poisson equation with Neumann boundary conditions. We use the centerlines and cross sections (Figure 2) to find mean values for the measured fluxes in the ascending and descending aorta, and in the supra-aortic arteries. These spatially averaged values of the measured aortic fluxes can be used as constraint fluxes, to be used as a base for corrections applied on the domain boundaries.

Relative Pressure Computation. To find the relative pressure map we have to solve the PPE equation with Neumann boundary conditions inside an irregular domain, which rewrite below in kinematic form (ν is the kinematic viscosity coefficient).

$$\frac{1}{\rho}\Delta p = \nabla \cdot RHS, \quad \frac{1}{\rho}\frac{\partial p}{\partial n} = RHS \cdot \hat{n} \tag{4}$$

$$RHS = -(\mathbf{u}_t + (\mathbf{u} \cdot \nabla)\mathbf{u}) + \nu\Delta\mathbf{u} \tag{5}$$

Let us consider the time-varying simply-connected domain D^n, with a locally Lipschitz interface C. We seek to solve the PPE on the domain D, and represent it by a level function Φ such that $D = \{x|\Phi(x) \leq 0\}$, $D^{outside} = \{x|\Phi(x) > 0\}$

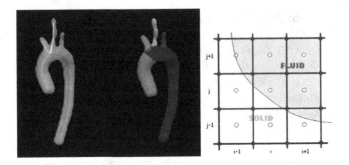

Fig. 2. Semi-automatic centerline tagging (left) and segmental division for regional mean flow computation (middle). Discretization of the variables: cell centered velocities and pressures, nodal level set.

and $C = \{x | \Phi(x) = 0\}$. We use a constant density, finite volume discretization of the equations on a rectangular (possibly non-isotropic) grid configuration in which the velocity and pressure are given at cell centers, while the level set is given at the nodes (Figure 2). The level set allows a first order accurate computation for the face fractions used in the discretization of the Laplace operator. In contrast, a staircase approach would have zeroth order accuracy. The transient inertia terms at an intermediate time step $n + 1/2$ are computed using temporal differences on the reference domain D^n at time n, onto which we warp the velocities from time step $n + 1$ using the displacement fields computed by the 4D Flow Tool. The velocity Laplacian and the divergence operators are discretized using centered differences.

By discretizing the PPE inside a computational cell C_{ij}, after using the divergence theorem and performing numerical manipulations, we obtain (we write this in 2D for simplicity):

$$l_{i-\frac{1}{2},j} \cdot \frac{p_{i,j} - p_{i-1,j}}{\Delta x} + l_{i+\frac{1}{2},j} \cdot \frac{p_{i,j} - p_{i+1,j}}{\Delta x} + l_{i,j-\frac{1}{2}} \cdot \frac{p_{i,j} - p_{i,j-1}}{\Delta y} + l_{i,j+\frac{1}{2}} \cdot \frac{p_{i,j} - p_{i,j+1}}{\Delta y} \tag{6}$$

$$= l_{i-\frac{1}{2},j} \cdot R^1_{i-\frac{1}{2},j} - l_{i+\frac{1}{2},j} \cdot R^1_{i+\frac{1}{2},j} + l_{i,j-\frac{1}{2}} \cdot R^2_{i,j-\frac{1}{2}} - l_{i,j+\frac{1}{2}} \cdot R^2_{i,j+\frac{1}{2}} \tag{7}$$

In the above we denoted $RHS = (R^1, R^2)$, and $l_{i,j}$ are the length (in 2D, face in 3D) fractions, which can be obtained with the help of the level set. Using the above discretization one puts together a symmetric positive definite linear system for the relative pressure, which is then solved iteratively using a multigrid method. Note that for interior nodes (defined as nodes with only fluid node neighbors, hence with face weights equal to one) one obtains the usual seven-point discretization of the 3D Laplace operator. For computational efficiency the numerical domain is tagged to include only the masked cells and their immediate outside neighbors and use a sparse matrix representation for the discrete Laplacian. This ensures a computation time of several seconds on single CPU for the 2.5mm resolution 4D flow MRI data sets that were considered, which is essential for fast clinical feedback.

4   Results and Discussion

We tested our methodology on both synthetic data and 4D flow MRI data. Each PPE computation took less than a minute for the full sequence 4D flow data, on a single Intel Xeon CPU with 2.53GHz frequency and a 32 bit machine.

The simplest in-silico test used analytical data for Poiseuille flow in a cylindrical pipe and our computed relative pressures matched the theoretical values with less than 0.3% error. The code was also tested for convergence on analytical tests and achieved between first and second order of accuracy in L_∞ norm.

The in-vivo measurements included 4D flow MRI in 4 healthy volunteers (2 male, 2 female) on a 3T MR system (3T, Magnetom TRIO, Siemens, Erlangen, Germany). All examinations were performed using a standard 12-element torso coil. 4D flow MRI consisted of a previously described, k-space segmented, rf-spoiled gradient echo sequence with interleaved 3-directional velocity encoding ([11]). Other imaging parameters were: TE=2.4 ms, TR=4.8 ms, flip angle=7°, field of view (FOV) = 320x240 mm, spatial resolution = 2.5x2.5x2.8 mm^3, temporal resolution = 38.4 ms, scan time 15-25 min, parallel imaging with reduction factor R=2. MRI acquisitions were synchronized to the heart and breathing cycle using prospective ECG-gating and adaptive diaphragm navigator gating. Data were acquired in a sagittal oblique 3D volume. The total scan time for the flow-sensitive measurement was 15-20 min (heart rate dependent). The data was processed using the 4D Flow tool and the flow-aware geometric cutting and tagging procedures previously described. After following the workflow presented

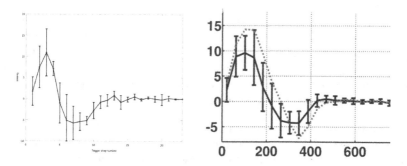

Fig. 3. Left: mean and standard deviation of pressure drop variation over time (mean inlet/AAo minus mean outlet/DAo) for the four healthy volunteers. The trigger step is 38 ± 1 milliseconds. Right: figure 4, last image, from [1].

in Section 3, relative pressure maps were obtained for each of the four healthy volunteers, and, for convenience of comparative analysis, the descending aorta mean outlet pressure was chosen as reference and the curves were scaled by the respective luminal volumes. We show in Figure 3 a very good match of the mean and standard deviation pressure drop between the AAo (mean over inlet as measured in this paper) and DAo (mean over outlet), as well as the variation of the pressure drop between the peak and its inversion peak, with the results obtained by [1] for the set of healthy volunteers. The peak-systolic pressure drop in the 4 volunteers along the centerline, between the AAo and DAo stations described above, varied between 5-10mmHg, similarly to the results obtained by [5].

We note furthermore that our relative pressure computation recovers the physiological inversion of the aortic pressure profile during diastole, associated with deceleration of blood flow during diastole as well as with the pressure pulse wave reflection at the periphery. As shown by [1], this inversion cannot be captured by using a Bernoulli approach to pressure estimation. A second qualitative observation is the variation of the computed pressure profile across the aortic lumen, especially in curved regions. This may be important to functionally assess clinically significant cases like bicuspid aortic valve (BAV), for which ascending aortic dilation and spatial pressure distribution and magnitude are correlated.

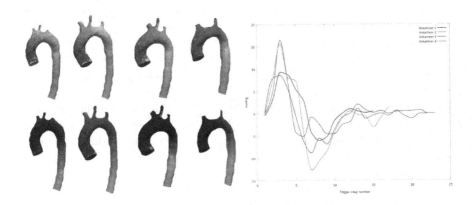

Fig. 4. Pressure drop variation over time (mean inlet minus mean outlet) for the four healthy volunteers. The top images show the pressure drops corresponding to the respective temporal peaks, while the bottom images show the pressure drops corresponding to the respective temporal troughs.

The success of our approach depends on the flow MRI quality: insufficient spatial and temporal data resolution may impede segmenting the supraaortic vessels, in which case one would need to use models for supraaortic flow. Testing will need to be done on more data sets, both healthy and diseased, to further validate the method and understand its limitations. Future work will also be focused on data assimilation techniques like Least-Squares FEM [4], which will use the whole time resolved 3D PC-MRI data prepared with the presented workflow, to constrain at each luminal grid point full 3D CFD computations. Furthermore, one can envision a direct application of the workflow for efficient computation of relative pressure, as a tool for assessment of risk in pathologies like aortic coarctation or BAV.

References

1. Bock, J., Frydrychowicz, A., Lorenz, R., Hirtler, D., Barker, A.J., Johnson, K.M., Arnold, R., Burkhardt, H., Hennig, J., Markl, M.: In vivo noninvasive 4D pressure difference mapping in the human aorta: Phantom comparison and application in healthy volunteers and patients. Magnetic Resonance in Medicine 66(4), 1079–1088 (2011)
2. Busch, J., Giese, D., Wissmann, L., Kozerke, S.: Reconstruction of divergence-free velocity fields from cine 3d phase-contrast flow measurements. Magnetic Resonance in Medicine 69(1), 200–210 (2013)
3. Currie, P.J., Seward, J.B., Reeder, G.S., Vlietstra, R.E., Bresnahan, D.R., Bresnahan, J.F., Smith, H.C., Hagler, D.J., Tajik, A.J.: Continuous-wave Doppler echocardiographic assessment of severity of calcific aortic stenosis: a simultaneous Doppler-catheter correlative study in 100 adult patients. Circulation 71(6), 1162–1169 (1985)
4. Dwight, R.P.: Bayesian inference for data assimilation using least-squares finite element methods. In: IOP Conf. Mat. Sci. Eng. (2010)
5. Ebbers, T., Farnebck, G.: Improving computation of cardiovascular relative pressure fields from velocity mri. Journal of Magnetic Resonance Imaging 30(1), 54–61 (2009)
6. Ebbers, T., Wigstrm, L., Bolger, A.F., Engvall, J., Karlsson, M.: Estimation of relative cardiovascular pressures using time-resolved three-dimensional phase contrast mri. Magnetic Resonance in Medicine 45(5), 872–879 (2001)
7. Fung, Y.: Biomechanics: Circulation. Springer (2010)
8. Gulsun, M.A., Jolly, M.P., Guehring, J., Guetter, C., Littmann, A., Greiser, A., Markl, M., Stalder, A.: A novel 4D flow tool for comprehensive blood flow analysis. In: Proceedings of ISMRM (2012)
9. Itu, L., Sharma, P., Gulsun, M., Mihalef, V., Kamen, A., Greiser, A.: Determination of time-varying pressure field from phase contrast MRI data. Journal of Cardiovascular Magnetic Resonance 14(suppl. 1), W36 (2012)
10. Krittian, S.B., Lamata, P., Michler, C., Nordsletten, D.A., Bock, J., Bradley, C.P., Pitcher, A., Kilner, P.J., Markl, M., Smith, N.P.: A finite-element approach to the direct computation of relative cardiovascular pressure from time-resolved MR velocity data. Medical Image Analysis 16(5), 1029–1037 (2012)
11. Markl, M., Harloff, A., Bley, T.A., Zaitsev, M., Jung, B., Weigang, E., Langer, M., Hennig, J., Frydrychowicz, A.: Time-resolved 3D MR velocity mapping at 3T: Improved navigator-gated assessment of vascular anatomy and blood flow. Journal of Magnetic Resonance Imaging 25(4), 824–831 (2007)
12. Markl, M., Kilner, P., Ebbers, T.: Comprehensive 4D velocity mapping of the heart and great vessels by cardiovascular magnetic resonance. Journal of Cardiovascular Magnetic Resonance 13(1), 7 (2011)
13. Meier, S., Hennemuth, A., Friman, O., Bock, J., Markl, M., Preusser, T.: Noninvasive 4d blood flow and pressure quantification in central blood vessels via pc-mri. In: Computing in Cardiology, pp. 903–906 (2010)
14. Tafti, P.D., Delgado-Gonzalo, R., Stalder, A.F., Unser, M.: Variational enhancement and denoising of flow field images. In: 2011 IEEE International Symposium on Biomedical Imaging: From Nano to Macro, March 30-April 2, pp. 1061–1064 (2011)
15. Tyszka, J.M., Laidlaw, D.H., Asa, J.W., Silverman, J.M.: Three-dimensional, time-resolved (4d) relative pressure mapping using magnetic resonance imaging. Journal of Magnetic Resonance Imaging 12(2), 321–329 (2000)

Self Stabilization of Image Attributes
for Left Ventricle Segmentation

Sarada Prasad Dakua[1,*], Julien Abi-Nahed[1], and Abdulla Al-Ansari[2]

[1] Qatar Robotic Surgery Centre, Qatar Science & Technology Park,
Qatar Foundation, Qatar
[2] Hamad Medical Corporation, Qatar
sdakua@qstp.org.qa

Abstract. Clinically, segmentation has many benefits for effective patient management, both in terms of pre-operative planning and post-operative assessment. Volumetric image segmentation of medical data still remains as a major challenge, largely due to the complexities of *in-vivo* anatomical structures, cross-subject and cross-modality variations. This correspondence presents a semiautomatic segmentation algorithm that is based on graph and chaos theory. Also, we introduce a new weighting function in the method for accurate delineation of regions of interest in medical images that contain regional inhomogeneities; the preliminary results show the potential of the proposed technique.

Keywords: Chaos concept, magnetic resonance images, segmentation.

1 Introduction

Image segmentation is known to provide adequate information about the shape and size of an object, therefore, it bears the utmost importance before any surgery. A rich tradition of work (for example, [1]) in image segmentation has focused on the establishment of appropriate image (object) models; because of the space constraint, we restrict ourselves only to a few related methods. Though fully automatic segmentation techniques are being pursued as a major research effort in the medical image computing community, the reliability of automatic methods is still inferior [1]. Without doubt, graph-based methods have advanced our understanding of image segmentation and have successfully been employed since sometime without heavy reliance on explicitly learned/encoded priors. Intelligent scissors is a boundary-based interactive method, that computes minimum-cost path between user-specified boundary points. However, this is unable to integrate any regional bias naturally, which is overcome by the Graph Cut method as follows. Graph Cut [2] is a combinatorial optimization technique; the globally optimal pixel labeling can be efficiently computed by maxflow/min-cut algorithms. Grab Cut [14] extends Graph Cut (GC) by introducing iterative segmentation scheme. However, current graph-cuts methods do exhibit certain

---

* Corresponding author.

O. Camara et al. (Eds.): STACOM 2013, LNCS 8330, pp. 244–252, 2014.

limitations. For example, in case of shape priors, the energy term usually contains additive shape energies which require templates or assume circular or ellipsoidal regions [3]. These priors greatly improve performance in the case of object classes with similar shapes or in the presence of templates or statistical shape models. However, the smoothness energy term in most graph-cuts methods is based on pixel intensities only. The pixel intensities can be locally erroneous due to noise and other image acquisition problems especially in medical data. The failure of these methods on medical images mostly attribute to the presence of high noise level and poor pixel intensity distribution in the images. In this scenario, we present a chaos based semi-automatic segmentation algorithm for left ventricle (LV) magnetic resonance (MR) images that exhibits graph theory. Here, we assume various objects in the image as vacillant which are aimed to stabilize and segment by applying the chaos concept; the preliminary results show the potential of the proposed technique.

2 Method and Materials

A French mathematician Henri Poincare first developed this chaotic model (CM) [4] by observing a significant deviation in the output if the input is varied even slightly. Edward Lorenz revisited this behavior with a set of 3 equations as follows:

$$dx/_{dt} = \sigma\left(y - x\right), \; dy/_{dt} = \left(rx - y - xz\right), \; dz/_{dt} = \left(xy - bz\right) \qquad (1)$$

where t, (x, y, z), σ, r and b denote time, dependent variables, Prandtl number, Rayleigh number and width-height ratio, respectively. Lorenz has found the values of σ, r, and b as 10, 25, and $\frac{8}{3}$, respectively as the best representation in his experiment [4]; presently, we too persist with these values.

2.1 Exposition of Chaos Concept for Image Segmentation

Our proposed method for image segmentation is based on graph theory where the candidate image is treated as a graph (G) or network; all edges in the graph are assigned some nonnegative weight (cost) by a weighting function, such as $w_e = e^{-\xi(x_i - x_j)^2}$ [5], where x_i is the pixel intensity at node i. The graph is decomposed into maximal strongly connected components (vertices of the graph). We look at the strongly connected components (largely responsible for segmentation) corresponding to vertices in the graph and call them the leading strong connected components (LSCC). Since there are usually more than one such LSCC due to the in-homogeneity nature of a typical medical image, it is not be possible to synchronize the entire network. Here the individual systems, meaning various objects present in an image, are considered chaotic. Therefore, in order to make a specific individual system stable (since we are interested in one object, say LV), it is required to impose some boundary conditions by placing seed points on the image. Here comes the chaotic theory that can be applied in image segmentation ensuring

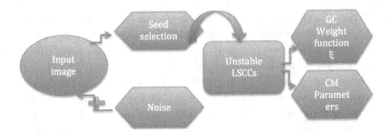

Fig. 1. Flow chart showing determination of ξ from CM parameters

deterministic convergence by keeping initial conditions constant. The scenario is analogous to- "iron particles are moving randomly in a cell and a strong magnet is suddenly placed on its center". Our objective is to find the probabilities of the particles reaching the magnet.

The stable points (σ, r and b) are responsible to stabilize the system; additionally, these are also useful to empirically determine ξ of w_e by following $\xi = \kappa + \frac{0.67r^{0.25}}{[1+1.48\sigma^{0.56}]^{\frac{4}{9}}}$ [6], κ being the mean intensity value of the image (as shown in Fig. 1). The probability of a particle to reach the magnet due to its magnetic force depends on its position, resistance along the trajectory and other constraints. In context with image, only one variable (gray-scale/pixel intensity value) is being considered; upon substituting the corresponding values for σ, r and b, the above equation can be generalized into a Dirichlet format [7] as $D[v] = \int_\Omega |\nabla v|^2 d\Omega$, where v is the field (twice differentiable boundary condition on $d\Omega$), and Ω is the region (domain Ω of $\mathbb{R}^n$). The solution of $D[v]$ has to satisfy the boundary conditions imposed by the initial seeds for image segmentation. In context with segmentation, image pixels are treated as iron particles and the probabilities with which a pixel reaches the initial seed points need to be found out.

2.2 Image Segmentation

The input image I is represented as a graph, $G = (V, E)$; where V and E represent the set of vertices and edges, respectively; pixels are the nodes of the graph. A weight is associated with each edge based on some property of the pixels that it connects, such as their image intensities.

- Input: The seed points are set on different labels in the image depending upon number of objects to be segmented.
- Laplacian matrix L is built based on the computed edge weights w.
- The Laplacian matrix is partitioned, $L = \begin{pmatrix} L_m & B \\ B^T & L_u \end{pmatrix}$. Subscript m and u represent likelihood for marked (seeded) and unmarked (unseeded) pixels, respectively.

- The linear system is set up as, $L_u X_u = -B^T X_m$
  where the variable X_u represents the set of probabilities corresponding to unseeded nodes, X_m is the set of probabilities corresponding to seeded nodes. L_u, B, and L_m correspond to the matrix decomposition of L.
- The linear system needs to be solved to get X_u (set of probabilities for each seed point).
- The most probable pixels are then marked with label numbers.
- Finally, computation of gradient on the image results non zero values at the object boundary.

2.3   Influence of Weighting Function

The weighting function is responsible to measure the similarity between two connected vertices and the performance of any graph based segmentation method depends on the choice of this function [8]. Usually, Gaussian function is a regular candidate for this purpose. However, the performance of Gaussian weighting function is limited on medical images due to high noise level and poor intensity distribution. Therefore, we explore for another suitable function that could improve the performance of Laplacian operator at the fundamental level and produce good segmentation results on blurry edged CMR images.

2.4   Derivative of Gaussian (DroG) Weighting Function

A blurred edge $v(x)$ is represented by a concatenation of an exponential function followed by a step function. Therefore, $v(x) = \begin{cases} e^{\left(-\frac{(x-t)^2}{2\sigma_s^2}\right)}, & x \le t \text{ , where } t \\ 1, & x > t, \end{cases}$
and σ_s denote the width of the region of interest and the extent of blurring of the blurred edge, respectively. Let a characteristic function (CF) E_v^w be defined that determines the suitability of a weighting function (w) to a particular edge (v); lesser its magnitude, more suitable is the corresponding weighting function. The characteristic function (CF) is given by $E_v^w = \int_0^{2t} f(x)\, w(x)\, dx,$, where $f(x)$ and w represent the input signal and candidate weighting function, respectively. The derivative of the Gaussian function $(Gauss(x))$ is obtained by,

$$\frac{\partial^n}{\partial x^n} Gauss(x) = (-1)^n \frac{1}{\left(\sigma\sqrt{2}\right)^n} H_n\left(\frac{x}{\sigma\sqrt{2}}\right) Gauss(x) \qquad (2)$$

where n is the order of the derivative and $H_n(x)$ is the Hermite polynomial; in our case we take $n = 1$. Simplifying the above function we get, $w_{DroG}(x) = \frac{-x\xi}{2\sqrt{2\pi}\sigma^3} e^{-\frac{x^2}{2\sigma^2}}$. The difference between two CFs can be expressed as,

$$
\begin{aligned}
E_v^{DroG} - E_v^G = \frac{1}{2\sqrt{2\pi}\sigma^3} &\left[e^{\left(\frac{t(\ln\ t-1)-\left(\frac{1}{2\sigma^2}+\frac{1}{2\sigma_s^2}\right)\frac{t^3}{3}+\frac{t^2}{4\sigma_s^2}}{t^3 e^{\left(-\left(\frac{1}{2\sigma^2}+\frac{1}{2\sigma_s^2}\right)\right)t^2}} \right)} -0.25\sigma^2 \left(\Gamma\left(2,\frac{2t^2}{\sigma^2}\right) - \Gamma\left(2,\frac{t^2}{2\sigma^2}\right) \right) \right. \\
&-\frac{1}{\sqrt{2}\sigma} \left[\frac{0.5}{\left(\frac{1}{2\sigma^2}+\frac{1}{2\sigma_s^2}\right)^{\frac{3}{2}}} \left(\left(\frac{t}{2\sigma_s^2}\right)^2 - \left(\frac{1}{2\sigma_s^2+2\sigma^2}\right)\left(\frac{t^2}{2\sigma_s^2}\right) \right) \left\{ erf\left(\sqrt{\frac{1}{2\sigma^2}+\frac{1}{2\sigma_s^2}}t - \frac{\frac{t}{2\sigma_s^2}}{\sqrt{\frac{1}{2\sigma^2}+\frac{1}{2\sigma_s^2}}} \right) \right. \right. \\
&\left. \left. -erf\left(\frac{\frac{-t}{2\sigma_s^2}}{\sqrt{\frac{1}{2\sigma^2}+\frac{1}{2\sigma_s^2}}} \right) \right\} + \frac{\sigma}{\sqrt{2}}\left(erf\left(\sqrt{\frac{1}{2\sigma^2}}2t\right) - erf\left(\sqrt{\frac{1}{2\sigma^2}}t\right) \right) \right]
\end{aligned}
\tag{3}
$$

where erf is the error function and $\Gamma(a,x) = \int_z^\infty e^{-t}t^{a-1}dt$. In order to show the effect of blurring, σ_s is varied (as shown in Fig. 3(i)), keeping the value of σ constant. For $t >= 3$, the plot shows negative values of $E_v^{DroG} - E_v^{Gauss}$, which means the CF magnitude due to DroG is less.

Table 1. The employed metrics for quantitative evaluation

Measure	Definition
Hausdorff distance (HD)	Minimum distance between two sets of points
False positive ratio (FPR)	Fragment of pixels incorrectly segmented
False negative ratio (FNR)	Fragment of pixels incorrectly rejected
Mean error rate (MER)	(False positive + false negative)/total samples × 100
Intra-region (I_h) uniformity	index of homogeneity inside a region
Specificity (Spec)	True negative/(true negative + false positive)
Precision (Prec)	True positive/(True positive + false positive)
Accuracy (Acc)	(True positive + true negative)/total samples
Sensitivity (Sens)	True positive/(true positive/false negative)
Dice coefficient (DC)	Quantity of overlapping of two contours
Pratt's Figure of merit (FOM)	segmentation accuracy

2.5   Method Summary

Four-five slices (if 5, S_i, $i = 1, ..., 5$), depending on the intensity distribution, of a subject are first empirically selected; one apex, one basal and the rest from the mid slices with clear boundary. These slices are segmented using the proposed method considering the 2nd (smallest) eigenvector of the Laplacian matrix as the optimal cut. Initially, the image is treated as a graph; seed points on both foreground and background determine the probability map. The label map is built by considering the maximum of two probabilities at a node.

Finally, gradient operation on the label map determines the coordinates that carry nonzero value as the desired contour coordinates. Next, we generate segmentation on the rest of the slices of the subject by following a level set procedure to build the volumetric LV where we keep the record of track changes of a contour (of a slice) until it reaches the next selected slice contour. These recorded tracks are the intermediate contours between the two segmented slices (S_i and S_{i-1}). In this way, the volumetric LV is formed.

3 Results and Discussion

The suggested segmentation algorithm is implemented on two databases, one from a hospital (33 subjects) [9] and the other from MICCAI Challenge 2009 (30 subjects) [10]. Ground truth images were provided along with the datasets by the respective organizers. In [9], each patients image sequence consisted of exactly 20 frames and the number of slices acquired varied between 8-15; spacing-between-slices ranged between 6mm - 13mm. Each image slice consisted of 256×256 pixels with a pixel-spacing of 0.93mm - 1.64mm. The qualitative results are shown in Fig. 2; a nearly complete match may be observed if the segmented images are compared with the corresponding ground truth images. Also, we evaluate the segmentation performance by some standard measures as provided in Table 1. The corresponding values of first six in the table should be as minimum as possible where as that of the rest should be as maximum as possible for a good segmentation output meaning that the resulting contour closely approaches the ground truth. We have just included the analysis of 20 subjects (randomly selected) from each dataset in this paper due to page constraint and presented the average quantitative figures of the 20 subjects in Fig. 3 to let the reader feel the difference between the methods GC and CM. Also, we evaluate its segmentation accuracy by comparing the results of the proposed method with some similar standard methods that have reportedly overcome the possible limitations of state of the art segmentation methods; this is summarized in Table 2. The average variance for each measure for the methods in this table can be summarized as $0.4 \pm 0.2, 0.0 \pm 0.005, 0.0 \pm 0.002, 0.0 \pm 0.005, 0.0 \pm 0.0004, 0.0 \pm 0.006, 0.0 \pm 0.004,$ $0.0 \pm 0.005, 0.0 \pm 0.03, 0.5 \pm 0.3,$ and 0.0 ± 0.01. The proposed method seems to have performed better if we examine the results in Fig. 3(a) -3(h) and Table 2. On MICCAI dataset [10], the mean Dice metric (DM) and mean of mean absolute distance (MAD) are found to be 90.0 ± 1.3 and 2.0 ± 0.7, respectively. As reference, one of the best methods [11] of MICCAI 2009 challenge achieved a mean DM of 91 ± 0.4 and mean MAD of 2.96 ± 1.09. The average time taken for segmentation is 5 s per slice (without optimization) on MATLAB 7.5 on a PC with Pentium 4, 3 GHz dual core processor.

Table 2. Segmentation comparison with different methods

Method	HD	DC	FPR	FNR	Sens	Spec	Prec	Acc	MER	I_h	FOM
NCut [8]	3.5453	.8343	.0234	.0084	.8356	.8487	.8534	.8490	.8009	4.21	0.86
Kmeans [12]	3.7422	0.8135	.0256	.0092	.8206	.8829	.8419	.8232	.8670	4.24	0.83
GrowCut [13]	3.2581	.8768	.0193	.0078	.8910	.7813	.9012	.9134	.6937	3.98	0.87
GrabCut [14]	3.2409	.8623	.0211	.0079	.8780	.8321	.8879	.8869	.7156	4.03	0.92
Our method	**3.0132**	**.9191**	**.0184**	**.0072**	**.9412**	**.7532**	**.9481**	**.9574**	**.6199**	**3.86**	**0.97**

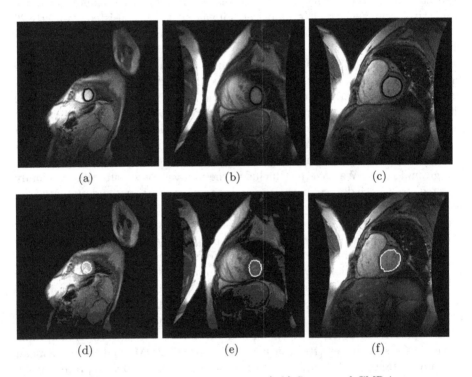

Fig. 2. (a-d) Ground truth CMR images. (e-h) Segmented CMR images

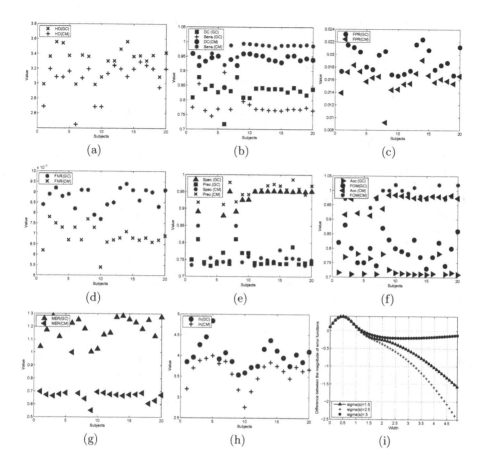

Fig. 3. Values of (a) HD. (b) DC and Sensitivity. (c) FPR. (d) FNR. (e) Specificity and Precision. (f) Accuracy and FOM. (g) MER. (h) I_h. (i) $E_v^{DroG} - E_v^G$.

4   Conclusions

This paper has presented a graph-based semi-automatic algorithm for volumetric LV reconstruction from a few slices based on Chaotic theory. Though the performance is satisfactory, the computational still remains a concern which we expect to reduce after its optimization. In future, we intend to explore its behavior on various subjects and different modalities.

References

1. Heimann, T., et al.: Comparison and evaluation of methods for liver segmentation from CT datasets. IEEE Trans. on Med. Imag. 28, 1251–1265 (2009)
2. Dakua, S., Abi-Nahed, J.: Patient oriented graph-based image segmentation. BSPC 8, 325–332 (2013)

3. Vu, N., Manjunath, B.: Shape prior segmentation of multiple objects with graph cuts. In: CVPR, pp. 1–8 (2008)
4. David, S.: The applicability principle: what chaos means for Social science. Behavioral Science 40, 22–40 (1995)
5. Grady, L.: Random walks for image segmentation. IEEE Trans. on Patt. Anal. and Mach. Intel. 28, 1–17 (2006)
6. Sijbers, J., Dekker, A.: Maximum likelihood estimation of signal amplitude and noise variance from MR data. Magn. Res. in Med. 51, 586–594 (2004)
7. Courant, R., Hilbert, D.: Methods of mathematical physics, vol. 2. Wiley and Sons, Berlin (1989)
8. Jianbo, S., Malik, J.: Normalized cuts and image segmentation. IEEE Trans. on Patt. Anal. and Mach. Intel. 22(8), 888–905 (2000)
9. Andreopoulos, A., Tsotsos, J.: Efficient and generalizable statistical models of shape and appearance for analysis of cardiac MRI. Medical Image Analysis 12, 335–357 (2008)
10. MICCAI Cardiac MR Left Ventricle Segmentation Challenge (2009)
11. Ben Ayed, I., Punithakumar, K., Li, S., Islam, A., Chong, J.: Left ventricle segmentation via graph cut distribution. In: Yang, G.-Z., Hawkes, D., Rueckert, D., Noble, A., Taylor, C. (eds.) MICCAI 2009, Part II. LNCS, vol. 5762, pp. 901–909. Springer, Heidelberg (2009)
12. Tong, Z., Nehorai, A., Porat, B.: K-means clustering-based data detection and symbol-timing recovery for burst-mode optical receiver. IEEE Trans. on Comm. 54, 1492–1501 (2006)
13. Vezhnevets, V., Konouchine, V.: Growcut - interactive multi-label n-d image segmentation by cellular automata. In: Proc. of Graphicon, pp. 150–156 (2005)
14. Shoudong, H., Wenbing, T., Desheng, W., Xue-Cheng, T., Xianglin, W.: Image segmentation based on Grab Cut framework integrating multiscale nonlinear structure tensor. IEEE TIP 18, 2289–2302 (2009)

A Framework for the Pre-clinical Validation of LBM-EP for the Planning and Guidance of Ventricular Tachycardia Ablation

Tommaso Mansi[1,*], Roy Beinart[2,3,4], Oliver Zettinig[1,5], Saikiran Rapaka[1], Bogdan Georgescu[1], Ali Kamen[1], Yoav Dori[6,7] M. Muz Zviman[2], Daniel A. Herzka[8], Henry R. Halperin[2,8,9], and Dorin Comaniciu[1]

[1] Imaging and Computer Vision, Siemens Corporate Technology, Princeton, NJ
[2] Cardiology Division, Johns Hopkins University School of Medicine, Baltimore, MD
[3] Davidai Arrhythmia Center, Leviev Heart Center, Sheba Medical Center, Ramat Gan, Israel
[4] Sackler School of Medicine, Tel Aviv University, Tel Aviv, Israel
[5] Computer Aided Medical Procedures, Technische Universität München, Germany
[6] Division of Cardiology, The Children's Hospital of Philadelphia, Philadelphia, PA
[7] Perelman School of Medicine at the University of Pennsylvania, Philadelphia, PA
[8] Department of Biomedical Engineering, Johns Hopkins University School of Medicine, Baltimore, MD
[9] The Russell H. Morgan Department of Radiology and Radiological Sciences, Johns Hopkins University School of Medicine, Baltimore, MD
tommaso.mansi@siemens.com

Abstract. This manuscript presents a framework for the pre-clinical validation of LBM-EP, a fast cardiac electrophysiology model based on the lattice-Boltzmann method (LBM). The overarching goal is to assess whether the model is able to predict ventricular tachycardia (VT) induction given lead location and stimulation protocol. First, the random-walk algorithm is used to interactively segment the heart ventricles from delayed-enhancement magnetic resonance images (DE-MRI). Scar and border zone are visually delineated using image thresholding. Then, a detailed anatomical model is generated, comprising fiber architecture and spatial distribution of action potential duration. That information is rasterized to a Cartesian grid, and the cardiac potentials are computed. The framework is illustrated on one swine data, for which two different pacing protocols at four different sites were tested. Each of the protocols were then virtually tested by computing seven seconds of heart beat. Model predictions in terms of VT induction were compared with what was observed in the animal. Our parallel implementation on graphics processing units required a total computation time of about two minutes at an isotropic grid resolution of $0.8\,mm$ (21s at a resolution of $1.5\,mm$), thus enabling interactive VT testing.

---

* Corresponding author.

O. Camara et al. (Eds.): STACOM 2013, LNCS 8330, pp. 253–261, 2014.
© Springer-Verlag Berlin Heidelberg 2014

1   Introduction

Ventricular tachycardia (VT) and ventricular fibrillation are among the most life-threatening cardiac events, causing the majority of the 200,000 yearly sudden deaths. In addition to anti-arrhythmic drugs, radio-frequency ablation constitutes a relatively efficient and cost-effective therapy. However, ablation in the setting of healed myocardial infarction has only 58% initial success rate and 71% eventual success rate following repeated procedures [1]. Possible explanations include: 1) the reentrant pathways to treat are complex and their origination point is challenging to map; 2) ablation is performed by successive, localized burns, which may not provide continuous lines of block; and 3) registration errors between electrophysiological mapping and anatomy make ablation planning and guidance inaccurate [5]. These limitations not only limit the success of the procedure but also significantly prolong the duration of the intervention and increase complication rates. There is therefore a need for new approaches to assist VT ablation therapy both prior and during the intervention to improve outcomes.

To tackle this challenge, computational models of cardiac electrophysiology (EP) are being investigated. In [13,15], preoperative magnetic resonance images (MRI) were employed to generate a model of patient's heart anatomy. Cardiac EP was then computed using patient-specific parameters estimated from intraoperative endocardial mapping at sinus rhythm [13]. The authors then virtually stimulated the myocardium close to the scar to induce VT. However, in both studies, the results were not validated against clinical observations. Pre-clinical studies have recently been performed to thoroughly evaluate the approach. In [11], a phenomenological model was employed to compute cardiac EP in two swine hearts, for which in-vivo EP measurements and ex-vivo diffusion tensor imaging (DTI) were available. The model was able to predict VT induction, using both DTI and synthetic fibers. In [10], the authors investigated whether an isotropic EP model, with generic parameters (i.e. non subject-specific), was able to predict VT induction. One protocol and different lead positions were virtually tested on eight pigs, showing promising predictions of VT induction compared to the observed inducibiliy. Yet, the question of whether the model was able to predict induction for a specific lead location and protocol was not tackled. A similar hypothesis has also been evaluated in 12 patients [2]. Promising results were obtained. VT induction could be predicted by the model in nine cases, but the electrocardiogram (ECG) morphology of VT and re-entrant circuits could not be validated. Hence, comprehensive validation on clinical setups is still missing.

As a first step towards in-silico testing of tachycardia induction, we propose a framework for the pre-clinical validation of a fast EP model, LBM-EP [12], in terms of VT planning. Combined with advanced image analysis methods (Sec. 2), our framework allows the fast computation of patient-specific EP for interactive virtual electrophysiology studies. Furthermore, an efficient ECG model is incorporated in the model to compute the predicted ECG VT morphology. Sec. 3 reports preliminary results of two pacing protocols at four different locations in one swine, for which VT inducibility is known. Sec. 4 concludes the manuscript.

2   Material and Methods

2.1   Pre-clinical Protocol

Animal Model. One swine was used in this study. Under general anesthesia, the middle-left anterior descending coronary artery was occluded between the first and second diagonal branch for 120 minutes using a $2.7\,Fr$ balloon angioplasty catheter via a femoral artery, to create a myocardial infarction (MI). Sixteen weeks after MI induction, the swine underwent in-vivo MRI and two days after the MRI, an EP study was done to determine inducibility of sustained VT.

Electrophysiological Evaluation. Detailed left ventricular (LV) mapping during sinus rhythm was performed with a multi-electrode 2-mm-tip catheter (6 Fr) with 2, 6, 2-mm inter-electrode spacing (Dynamic XT, Bard Electrophysiology, Lowell) to construct a 3D voltage map using an electro-anatomic mapping system (NavX, St. Jude Medical). Peak-to-peak bipolar amplitudes were displayed color-coded and recorded, with electrograms $\leq 1.5\,mV$ defined as low voltage electrograms. The programmed electrical stimulation protocol was conducted to induce VT using a pacing catheter (6 Fr) advanced to the right and then left ventricular chambers through a femoral vein. The stimulation protocol consisted of three decreasing extra-stimuli at two different drive cycle lengths (Sec. 3.2).

MRI Protocol. Imaging took place on a 3.0 Tesla system with a 32-channel cardiac phased array (Achieva, Philips Medical Systems, Best, The Netherlands). Global cardiac function was measured using 2D breath-hold balanced steady-state free precession ($1.25 \times 1.25 \times 5.0\,mm^3$). For visualization of the border zone (BZ) and scar, a custom 3D delayed contrast enhancement (DE-MRI) sequence (ECG-gated, respiratory navigator gated, phase-sensitive inversion recovery spoiled gradient echo [8]) with the following imaging parameters was used: 60 slices with 1.00x1.25x3.0 mm³ in-plane resolution reconstructed to $0.75 \times 0.75 \times 1.5\,mm^3$, TR/TE 5.6/2.7 ms, 18° flip angle, acquired $\approx$ 25-35 min post intravenous administration of $0.2\,mmol/kg$ Magnevist (Berlex/Schering AG, Berlin, Germany).

2.2   Model of Cardiac Anatomy

The bi-ventricular geometrical model was segmented from the DE-MRI using an interactive method based on the random-walk algorithm [7] (the atria were not considered in this study). Scar and BZ regions were delineated using image thresholding. In particular, a threshold of approximatively half the grey level peak intensity of the myocardium was used (image artifacts were removed from the calculation of the myocardium intensities). A second threshold was visually selected for the BZ to cover the hyper-intense area around the scar but not the healthy tissue. The resulting segmentations were fused to form a closed surface of the biventricular myocardium (Fig. 1, left panel) and mapped back onto a Cartesian grid using a level-set representation.

To cope with tissue anisotropy, a model of fiber architecture was calculated by following a rule-based approach [12]. Below the basal plane, fiber elevation angle varied linearly from the epicardium to the endocardium (from $-60°$ to $+60°$ for the LV, from $-80°$ to $+80°$ for the right ventricle (RV)), which were then geodesically extrapolated up to the valves (Fig. 1, right panel). Finally, a model of the spatial heterogeneity of the action potential distribution (APD) was used by spatially varying the parameter τ_{close} of the EP model (Sec. 2.3). In addition to the endocardial and epicardial cells ($\tau_{close_{endo}}$ and $\tau_{close_{epi}}$ respectively), we incorporated in our model the M-cells ($\tau_{close_{mid}}$) as it has been showed they contribute to the T-wave morphology [3]. A slight base-to-apex gradient (base APD equal to 95% of apex APD) was also used (Fig. 1, right panel)).

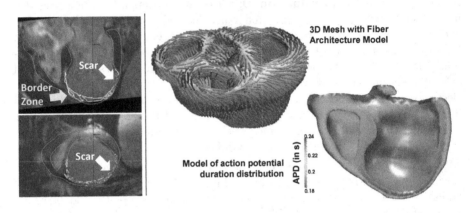

Fig. 1. Subject-specific anatomical model. *Left*: Heart segmentation (scar in red). *Right*: Mesh visualization of the anatomical model estimated from the segmentation.

2.3 Model of Cardiac Electrophysiology

The LBM-EP method was used to compute cardiac EP in a computationally efficient way. LBM-EP solves any mono-domain model by using the Lattice-Boltzmann method (LBM) [12]. In this work, the trans-membrane potential $v(\mathbf{x}, t) \in [-70\,mV, 30\,mV]$ was calculated according to the Mitchell-Schaeffer model [9]:

$$\frac{\partial v(\mathbf{x}, t)}{\partial t} = h\frac{v^2(1 - v)}{\tau_{in}} - \frac{v}{\tau_{out}} + c\nabla \cdot \mathbf{D}\nabla v + I_{stim}$$

$$\frac{dh(\mathbf{x}, t)}{dt} = \begin{cases} (1 - h)/\tau_{open} & \text{if } v < v_{gate} \\ -h/\tau_{close} & \text{otherwise} \end{cases}$$

where, $\mathbf{x}$ is the spatial location, t is the time, $h(\mathbf{x}, t)$ is a gating variable that models the state of the ion channels, c is the tissue diffusivity whose anisotropy is captured by the tensor $\mathbf{D}$ and I_{stim} is the stimulation current at the probe position. The parameters τ and v_{gate} control the dynamics of the action potential.

The mono-domain equation was solved on a 7-connectivity Cartesian grid (six edges and central position). For higher spatial accuracy, Neumann boundary conditions were enforced by using a level-set representation of the heart anatomy. Stimulation currents were applied through Dirichlet boundary conditions at the position of the pacing lead. Seven different domains were identified on the Cartesian grid: the left and right ventricular septum, used to pace the heart and to mimic the His bundle; the left and right endocardia, with fast electrical diffusivity, c_{LV} and c_{RV}, to mimic the Purkinje network; the myocardium, with slower diffusivity c_{Myo}, the BZ (c_{bz}) and the scar, which does not conduct the electrical wave altogether.

We also computed the ECG resulting from the calculated EP. The extracellular potentials were obtained according to an algebraic equation [4] and mapped to the torso using the boundary element method. Finally, the algorithm was implemented on a graphics processing unit (GPU) for maximal performance.

2.4 Virtual Pacing Protocol

Virtual pacing was performed interactively. The user first chose the protocol to apply in terms of pacing interval. The pacing lead was then placed on the 3D mesh interactively. Finally, cardiac EP was computed over $7\,s$ to cover the pacing session. Stimulus pulse width was $2\,ms$.

3 Experiments and Results

3.1 Model Personalization

For all experiments, EP was computed on a $0.8 \times 0.8 \times 0.8\,mm^3$ grid. Tissue diffusivity and action potential duration (APD) were manually adjusted to match the measured QRS duration (QRSd $= 116\,ms$) and QT interval (QTd $= 488\,ms$) at sinus rhythm. Fig. 2 shows the depolarization time map and the I-lead ECG calculated by the model. After personalization, computed QRSd and QTd matched the measurements: $\mathrm{QRSd}_{comp.} = 117\,ms$ and $\mathrm{QTd}_{comp.} = 445\,ms$ ($c_{LV} = c_{RV} = 1500\,mm^2/s$, $c_{Myo} = 400\,mm^2/s$, $\tau_{close_{endo}} = 180\,ms$, $\tau_{close_{mid}} = 190\,ms$ and $\tau_{close_{epi}} = 140\,ms$). τ_{open} was globally increased to match trends reported in [13] ($\tau_{open} = 200\,ms$). The nominal values $\tau_{in} = 0.3\,ms$ and $\tau_{out} = 6\,ms$ were used. The diffusivity in the scar was set to $0\,mm^2/s$. BZ diffusivity was assumed to be half of the healthy tissue, $c_{BZ} = 200\,mm^2/s$ while τ_{close} was increased by $30\,ms$ to mimic the longer APD observed in the healing tissue [10].

3.2 Virtual Electrophysiological Evaluation

For all experiments, natural septal pacing occurred at $t = 80\,ms$, then every $1180\,ms$, the measured cycle length. The first stimulus (S1) was applied at $t = 600\,ms$, except when mentioned otherwise, followed by the subsequent stimulations according to the selected protocol. Two stimulation protocols (P1:

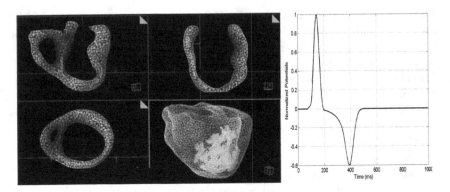

Fig. 2. Computed depolarization times (left) and I-lead ECG (right) at sinus rhythm

$8 \times 400\,ms$ (S1), $360\,ms$ (S2), $320\,ms$ (S3), $290\,ms$ (S4); P2: $8 \times 450\,ms$ (S1), $310\,ms$ (S2), $250\,ms$ (S3), $230\,ms$ (S4) were tested at four different locations (right ventricle endocardium apex (RV), right ventricle outflow tract (RVOT), left ventricle endocardium apex (LV), left ventricle outflow tract (LVOT), Fig. 3). For P2-LV, S1 was applied at $500\,ms$ in order to avoid the refractory period.

Fig. 4 reports the computed I-lead ECG traces. For all virtual EP evaluations, sustained VT could not be induced. This finding was consistent with what was observed in the animal for P1 protocols and P2-RV, P2-RVOT and P2-LVOT. For P2-LV, three monomorphic VT beats could eventually be observed in the animal but they immediately degenerated into fibrillation. The model, with these parameters, was not able to capture this pattern. Yet, non-reported experiments with an increased τ_{close} value inside the BZ yielded monomorphic VT, suggesting that more localized model personalization is necessary. Interestingly though, while for all P1 experiments no arrhythmias could be induced, the second protocol yielded some non sustained arrhythmias during or shortly after the pacing. Visual inspection of the dynamics of the trans-membrane potential identified that these arrhythmias happened when the electrical wave generated by the pacing lead collided with the one resulting from the natural pacing. It should be noted that because the stimuli were applied as Dirichlet boundary conditions (i.e. the trans-membrane potential $v(\mathbf{x}, t)$ was prescribed during the stimulus), no pacing artifacts due to the current source could be computed in the ECG. Fig. 4 therefore shows only the resulting ECG complexes.

3.3   Computational Efficiency

On a standard desktop machine (Intel Xeon 8-core @ 2.4 GHz, 4GB RAM, NVIDIA GeForce GTX 580), EP over $1\,s$ was computed in $19.5\,s$, yielding approximatively 2 minutes of computation for a complete VT pacing protocol. To the best of our knowledge, this is the first time interactive virtual VT study could be performed at relatively high resolution. It should be noted that on a $1.5\,mm$-isotropic grid, $3.4\,s$ only where needed to calculate $1\,s$ of EP. However,

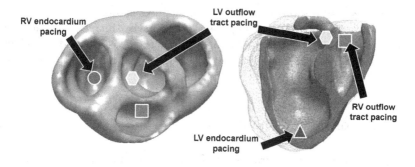

Fig. 3. Position of the leads for the four VT pacing protocols tested in this study

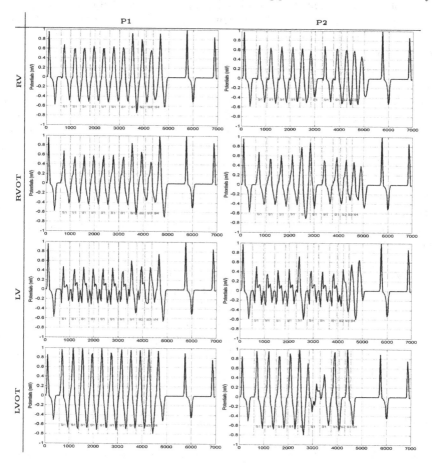

Fig. 4. Computed I-lead ECG for the different pacing protocols. None of them induced sustained VT after the end of the stimulations. Stimuli are represented by the red lines. See text for details.

using such a coarse resolution is not recommended as details of the scar and
BZ morphology would be lost compared to the native image resolution, thus
hindering the accuracy of the predictions. Further speed-up could be achieved
by using a more powerful graphics card, LBM-EP being highly scalable both in
terms of computational node and number of cores [6].

4   Discussion and Conclusion

This manuscript presented a framework for the pre-clinical validation of LBM-
EP, a fast computational model of cardiac electrophysiology, in terms of prediction of VT induction. Our approach couples advanced image analytics for
patient-specific anatomical modeling with a GPU-implementation of LBM-EP
and a model of ECG. Two minutes are needed to compute 7 s of heart beat, which
is enough to detect if VT can be induced or not, compared to the hours of computations needed in [2,10] (although a direct comparison is difficult due to the
different systems and methods). Our framework thus enables interactive virtual
electrophysiological evaluations. Moreover, contrary to current approaches, the
ECG morphology of VT could be computed, which will enable more thorough
evaluation of the framework. Preliminary experiments in one swine confirmed
the feasibility of the approach. Results were promising in terms of VT induction
prediction. More localized personalization would further improve prediction accuracy. Non-reported experiments with different fiber orientations (from −45°
at the epicardium to +60° at the endocardium as suggested in [11]) yielded the
same clinical conclusion: VT could not be virtually induced. In this work, we
included the M-cells into our model. Numerical simulations showed their importance on the T-wave morphology. Yet, while experimental studies on adult swine
confirmed the presence of this type of cells [14], they may not be present in ju-
venile swine [11]. More experimental and numerical studies would be needed to
evaluate this aspect of the model. The next steps consist in comprehensive and
quantitative evaluation against ECG signals and invasive endocardial mapping.
More cases are also being evaluated, along with a comprehensive sensitivity analysis of the model predictions with respect to the input parameters, in particular
conduction velocity, refractory period duration, scar and BZ segmentation, as
well as lead position and S1 starting time.

References

1. Aliot, E.M., Stevenson, W.G., Almendral-Garrote, J.M., Bogun, F., Calkins, C.H.,
   Delacretaz, E., Della Bella, P., Hindricks, G., Jaïs, P., Josephson, M.E., et al.:
   Ehra/hrs expert consensus on catheter ablation of ventricular arrhythmias developed in a partnership with the european heart rhythm association (ehra), a registered branch of the european society of cardiology (esc), and the heart rhythm
   society (hrs); in collaboration with the american college of cardiology (acc) and
   the american heart association (aha). Europace 11(6), 771–817 (2009)

2. Ashikaga, H., Arevalo, H., Vadakkumpadan, F., Blake III, R.C., Bayer, J.D., Nazarian, S., Muz Zviman, M., Tandri, H., Berger, R.D., Calkins, H., et al.: Feasibility of image-based simulation to estimate ablation target in human ventricular arrhythmia. Heart Rhythm (2013)
3. Boulakia, M., Cazeau, S., Fernández, M.A., Gerbeau, J.F., Zemzemi, N.: Mathematical modeling of electrocardiograms: a numerical study. Annals of Biomedical Engineering 38(3), 1071–1097 (2010)
4. Chhay, M., Coudière, Y., Turpault, R.: How to compute the extracellular potential in electrocardiology from an extended monodomain model. Research Report RR-7916, INRIA (March 2012)
5. De Bakker, J., Van Capelle, F., Janse, M., Wilde, A., Coronel, R., Becker, A., Dingemans, K., Van Hemel, N., Hauer, R.: Reentry as a cause of ventricular tachycardia in patients with chronic ischemic heart disease: electrophysiologic and anatomic correlation. Circulation 77(3), 589–606 (1988)
6. Georgescu, B., Rapaka, S., Mansi, T., Zettinig, O., Kamen, A.: Towards real-time cardiac electrophysiology computations using gp-gpu lattice-boltzmann method. In: High-Performance Computing, a 2013 MICCAI Workshop (2013)
7. Grady, L.: Random walks for image segmentation. IEEE Transactions on Pattern Analysis and Machine Intelligence 28(11), 1768–1783 (2006)
8. Lee, S.L., Schär, M., Kozerke, S., Harouni, A.A., Sena-Weltin, V., Zviman, M.M., Halperin, H., McVeigh, E.R., Herzka, D.A.: Independent respiratory navigators for improved 3d psir imaging of myocardial infarctions. Journal of Cardiovascular Magnetic Resonance 13(suppl. 1), P18 (2011)
9. Mitchell, C., Schaeffer, D.: A two-current model for the dynamics of cardiac membrane. Bulletin of Mathematical Biology 65(5), 767–793 (2003)
10. Ng, J., Jacobson, J.T., Ng, J.K., Gordon, D., Lee, D.C., Carr, J.C., Goldberger, J.J.: Virtual electrophysiological study in a 3-dimensional cardiac magnetic resonance imaging model of porcine myocardial infarction. Journal of the American College of Cardiology 60(5), 423–430 (2012)
11. Pop, M., et al.: A 3d mri-based cardiac computer model to study arrhythmia and its in-vivo experimental validation. In: Metaxas, D.N., Axel, L. (eds.) FIMH 2011. LNCS, vol. 6666, pp. 195–205. Springer, Heidelberg (2011)
12. Rapaka, S., Mansi, T., Georgescu, B., Pop, M., Wright, G.A., Kamen, A., Comaniciu, D.: LBM-EP: Lattice-boltzmann method for fast cardiac electrophysiology simulation from 3d images. In: Ayache, N., Delingette, H., Golland, P., Mori, K. (eds.) MICCAI 2012, Part II. LNCS, vol. 7511, pp. 33–40. Springer, Heidelberg (2012)
13. Relan, J., Chinchapatnam, P., Sermesant, M., Rhode, K., Ginks, M., Delingette, H., Rinaldi, C.A., Razavi, R., Ayache, N.: Coupled personalization of cardiac electrophysiology models for prediction of ischaemic ventricular tachycardia. J. R. Soc. Interface Focus 1(3), 396–407 (2011)
14. Stankovicova, T., Szilard, M., De Scheerder, I., Sipido, K.R.: M cells and transmural heterogeneity of action potential configuration in myocytes from the left ventricular wall of the pig heart. Cardiovascular Research 45(4), 952–960 (2000)
15. Talbot, H., Marchesseau, S., Duriez, C., Sermesant, M., Cotin, S., Delingette, H.: Towards an interactive electromechanical model of the heart. Interface Focus 3(2) (2013)

Image-Based Estimation of Myocardial Acceleration Using TDFFD: A Phantom Study

Ali Pashaei[1], Gemma Piella[1], Nicolas Duchateau[2], Luigi Gabrielli[2,3], and Oscar Camara[1]

[1] Universitat Pompeu Fabra, Barcelona, Spain
[2] Hospital Clínic, IDIBAPS, Universitat de Barcelona, Spain
[3] Pontificia Universidad Católica de Chile, Santiago, Chile

Abstract. In this paper, we propose to estimate myocardial acceleration using a temporal diffeomorphic free-form deformation (TDFFD) algorithm. The use of TDFFD has the advantage of providing B-spline parameterized velocities, thus temporally smooth, which is an asset for the computation of acceleration. The method is tested on 3D+t echocardiographic sequences from a realistic physical heart phantom, in which ground truth displacement is known in some regions. Peak endocardial acceleration (PEA) error was 20.4%, part of error being due to the low temporal resolution of the images. The allure of the acceleration profile was reasonably preserved. The study suggests a non-invasive technique to measure cardiac acceleration that may be used to improve the monitoring of cardiac mechanics and consecutive therapy planning.

1 Introduction

Several studies have shown that cardiac acceleration such as peak endocardial (PEA) and epicardial acceleration can be used as an index to assess function of the heart (such as max dp/dt) and detection of myocardial ischemia [1, 2]. For this reason, accelerometer devices have been proposed for monitoring myocardial ischemia and cardiac resynchronization therapy (CRT) optimization [2, 3, 4]. A pilot study on optimization of CRT has shown significant increase in the proportion of patients who improved with therapy [5]. Despite their advantages, accelerometer measurements are highly invasive, and only available at a single-location. Also, it was indicated that accelerometer measurements are influenced by patient orientation and gravity [6, 7].

Image-based computation of cardiac acceleration has been recently proposed, which may resolve these issues [8]. This study computed the acceleration directly from the displacement fields estimated by an image-based registration algorithm [9].

In the present work, we address this issue by computing acceleration directly from the parametric differentiation of myocardial velocities, the estimation of which is inherent to the registration process. The use of temporal diffeomorphic free-form deformation (TDFFD) algorithm has the advantage of providing B-spline parameterized velocities, thus spatiotemporally smooth, which is an asset

O. Camara et al. (Eds.): STACOM 2013, LNCS 8330, pp. 262–270, 2014.

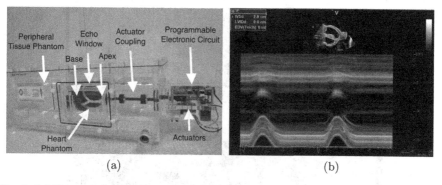

Fig. 1. (a) Dynamic heart phantom, and (b) M-mode data showing the sinus waveform displacement function of equation (1) generated by the phantom

for the computation of acceleration directly from these velocities. We extend the evaluation of the accuracy of such computations using data from a realistic physical heart phantom, for which ground truth is known at specific locations.

2   Materials and Methods

2.1   Imaging Data

A dynamic multimodality heart phantom (DHP-01, Shelley Medical Imaging Technologies, London, ON, CA) was employed to provide data for computation of the cardiac acceleration from images. This phantom mimics realistic anatomical geometry of the left and right ventricles of the human heart. The dynamic heart phantom setup is shown in Fig. 1(a).

This phantom has two controlable actuation systems for translational and rotational movement. Translation phantom actuator (located at the apex) was programmed to follow a sinus waveform, without any rotation. The function for the movement of actuator was:

$$\mathbf{d}_{apex} = \frac{A}{2}\left(1 - \cos(2\pi t/T)\right),\tag{1}$$

hence providing the acceleration as;

$$\mathbf{a}_{apex} = \frac{A}{2}\left(\frac{2\pi}{T}\right)^2 \cos(2\pi t/T).\tag{2}$$

The actuator was set to have stroke value $A = 0.02\,m$ within a period of $T = 1.05\,sec$. The movement contained a short delay after each movement, hence providing a heart rate of $34\,bpm$. The movement contained a short delay after each movement, hence providing a heart rate of $34\,bpm$.

Fig. 1(b) shows 2D M-mode visualization of the programmed phantom motion, for qualitative checking purposes.

3D+t echocardiographic sequences were acquired for the acceleration computations. Fig. 2 shows a representation of the acquired 3D data. Animated version

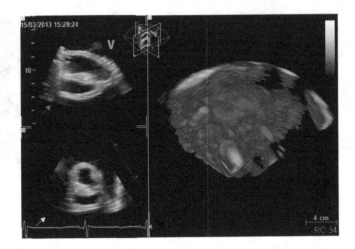

Fig. 2. Acquired 3D data at end-diastole

of this figure is available online here. The difference with real data acquisitions is that the phantom is surrounded by water inside and outside its walls, contrary to clinical images where tissue is visible outside the cardiac cavities, and therefore can make the image quality lower and the estimation of wall motion more difficult. Two temporal resolutions (11.9 and 21.7 fps) were used for 3D+t echocardiographic acquisition. All images were acquired with gating to a simulated vector ECG.

2.2 Image Processing

Segmentation of the left and right ventricles was first performed using 3D Active Shape Models (ASM) [10]. The mesh resulting from the segmentation was matched to the 3D+t echocardiographic sequences and propagated along the cycle by means of a non-rigid image registration algorithm (TDFFD) [9]. We used a two-level multiresolution implementation of the TDFFD. The initial grid size was of one control point per frame in the temporal direction, 5 control points in the short-axis direction and in the long-axis direction. As similarity measure, we used the sum of squared differences between the intensities of each frame and a reference (end-diastole) one. We used the L-BFGS-B as optimizer [11]. The left ventricle was divided using the 17-segment model as proposed by the American Heart Association (AHA) [12].

2.3 Computation of Acceleration

Acceleration is computed as the derivative of the estimated velocity function, which is defined in terms of B-Spline kernels. Here we modified this function to estimate acceleration directly from the TDFFD output. If the B-Spline coefficients of all contol points are presented in a vector of parameters $\mathbf{p}$, the velocity $\mathbf{v}(\mathbf{x}, t; \mathbf{p})$ is computed as

$$\mathbf{v}(\mathbf{x}, t; \mathbf{p}) = \sum_{i,j,k,l} \beta\left(\frac{x - x_i}{\Delta_x}\right) \beta\left(\frac{y - y_j}{\Delta_y}\right) \beta\left(\frac{z - z_k}{\Delta_z}\right) \beta\left(\frac{t - t_l}{\Delta_t}\right) \mathbf{p}_{i,j,k,l}, \quad (3)$$

where $\mathbf{x} = (x, y, z)$, $\beta(\cdot)$ is a 1D cubic B-Spline kernel, $\{x_i, y_j, z_k, t_l\}$ define a regular grid of 4D control points, and $\Delta_x, \Delta_y, \Delta_z, \Delta_t$ are the spacings between control points in each dimension. We can compute the acceleration $\mathbf{a}(\mathbf{x}, t; \mathbf{p})$ as

$$\mathbf{a}(\mathbf{x}, t; \mathbf{p}) = \frac{d\mathbf{v}(\mathbf{x}, t; \mathbf{p})}{dt} = \sum_{i,j,k,l} \beta\left(\frac{x - x_i}{\Delta_x}\right) \beta\left(\frac{y - y_j}{\Delta_y}\right) \beta\left(\frac{z - z_k}{\Delta_z}\right) \frac{d}{dt}\beta\left(\frac{t - t_l}{\Delta_t}\right) \mathbf{p}_{i,j,k,l},$$

$$(4)$$

Acceleration is computed on all nodes of the mesh obtained from segmentation of biventricular heart.

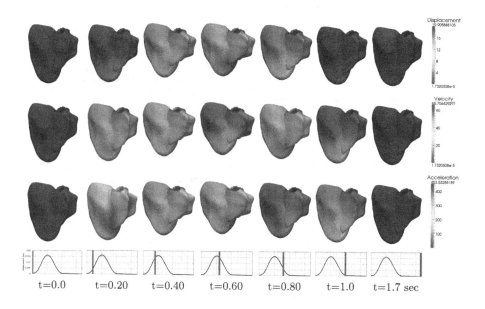

t=0.0   t=0.20   t=0.40   t=0.60   t=0.80   t=1.0   t=1.7 sec

Fig. 3. Distribution of the displacement (upper row), velocity (middle row) and acceleration (lower row) magnitudes during the heart cycle. Units for displacement, velocity and acceleration are m, m/s and m/s^2, respectively.

3 Results

3.1 Displacement, Velocity and Acceleration in Heart

Fig. 3 shows the magnitude of the estimated displacement, velocity and acceleration of the myocardium, along the cardiac cycle. These results are computed based on the image acquisition with frame rate 21.7 fps. Results were coherent with the echocardiographic images, and the program used for the phantom: stationary position in the cycle before $t = 0.20 \ sec$ and after $t = 1 \ sec$; higher values at the apex with respect to the basal level, and at end-diastole. Values change smoothly both in time and space due to the use of the TDFFD algorithm.

Results are detailed in Fig. 4 for four locations covering the myocardium from base to apex. The four points correspond to epicardium on apex, apical inferior, mid inferior and basal inferior section of the AHA 17-segment model, which are labeled with 17, 15, 10 and 4 respectively. Displacement and acceleration components in radial, circumferential and longitudinal directions are presented in Fig. 4(a) and (b), respectively. Realistic displacements are observed: sinus waveform as programmed along the radial direction at the apex (note that due to the phantom materials and setup, radial contraction is not necessarily guaranteed in the other regions); almost no circumferential motion; and decreasing longitudinal motion from apex to base.

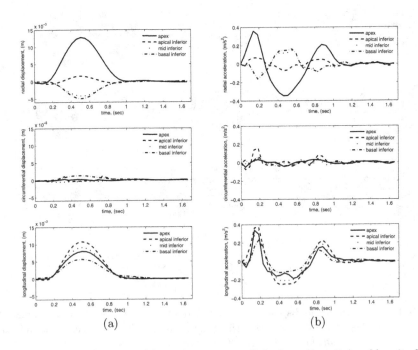

(a)   (b)

Fig. 4. (a) Displacement and (b) acceleration in radial, circumferential and longitudinal directions on four regions on epicardium during heart cycle

3.2 Displacement and Acceleration in Apex

Solid curve in Fig. 5(a) illustrates the recovered displacements at the apex from TDFFD algorithm. The presented curve is the average magnitude of displacement over the 17th AHA segment, along the cardiac cycle. Dashed line in this figure is the configured displacement function from equation (1). TDFFD follows the configured displacement function with mean squared error (MSE) of $5.15 \times 10^{-6} \, m^2$. The magnitude of peak displacement was $16.2 \, mm$, which corresponds to 19% relative error with respect to the settings of the phantom actuator.

The solid line in Fig. 5(b) shows the cardiac acceleration estimated by TDFFD at the apex. Dashed line in this figure shows the second analytic derivative of displacement function in equation (2). The MSE error from this estimation was found to be $0.011 \, m^2/s^4$. PEA was measured $0.285 \, m/s^2$, which shows 20.4% error with respect to the ground truth peak acceleration $0.358 \, m/s^2$ computed from equation (2).

3.3 Effect of Frame Rate

We performed an experiment where the frame rate was decreased from 21.7 to $11.9 \, fps$ to study the impact of temporal resolution on the image based estimations. Fig. 5(c) shows the radial acceleration at the same point in the apex computed from high and low frame rate imaging data. Acceleration magnitude has considerably changed under the effect of a lower temporal resolution. PEA from the low frame rate images was measured $0.235 \, m/s^2$, which under-estimates this value by factor of 34.3% and 17% with respect to the ground truth data and estimation from high frame rate images, respectively.

4 Discussion

Computing myocardial acceleration from medical images has several advantages over other existing techniques: the measurement is non-invasive and provides information at any location of the myocardium, and not at a single point. Furthermore (not demonstrated here) accelerometer sensors are sensitive to environmental factors such as gravity [7], which is not the case of image-based data.

In this study, we presented a method for estimating cardiac acceleration directly from temporal sequences of images. We used a non-rigid registration algorithm (TDFFD) to track the myocardium along the cycle, which has several advantages: (i) it provides differentiable velocities, necessary to the computation of acceleration; (ii) output data is smooth in both time and space, which may prevent from artifacts due to low image quality and low temporal resolution.

Fig. 5 shows the performance of this method for tracking the sinus waveform displacement. The error on peak displacement was 19% and PEA error was 20.4% and 34.3% using imaging data with frame rate 21.7 and $11.9 \, fps$, respectively. Some possible reasons for this error can be the sensitivity of the measurements to image quality and temporal resolution. This hypothesis is reinforced by our

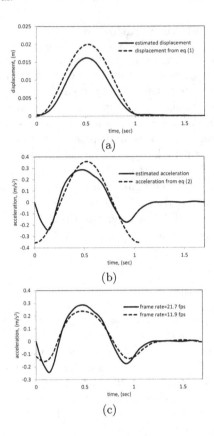

Fig. 5. (a) Estimated displacement (solid line) versus analytical plot of equation (1) (dashed lines), (b) estimated acceleration (solid line) compared to analytical plot of equation (2) (dashed lines), and (c) estimated acceleration from image data of frame rate 21.7 *fps* (solid line) comapred to that of image data with frame rate 11.9 *fps* (dashed line)

experiment on lower frame rate sequences, which results in a drop-off of the acceleration accuracy.

In our study, the acceleration was computed at every spatiotemporal location, but validation against ground truth was only possible at apex. The use of a FFD-based registration scheme, which intrinsically contains spline interpolation, may result in oversmoothing of the estimated curves. While this is not desirable for real applications measuring faster motion and deformation patterns, it also prevents the acceleration from being too noisy and therefore unexploitable. This is one limit of our study, that one may consider less relevant in the future with the availability of higher frame rates for 3D echocardiography.

Sonomicrometry cristals may be used to provide ground truth at other specific locations, but this was not retained in our study due to the risk of damage to the phantom material. Tagged-MRI is another option usually used as a surrogate of ground truth, but this would require a MR compatible phantom, which is not the case of the one we have.

5   Conclusion

We tested a method for estimating myocardial acceleration directly from image sequences, using non-rigid registration techniques (TDFFD), which has the advantage of providing spatiotemporally smooth velocities, and allows to compute acceleration directly from them. Experiments on a realistic phantom showed feasibility of recovery of acceleration profiles using this method. However peak acceleration measurement was conditioned by image quality and temporal resolution. It suggests a non-invasive technique to measuring the cardiac acceleration that could be used to improve the monitoring of cardiac mechanics and optimization of CRT.

Acknowledgements. This work has been partially supported by the Spanish Ministry of Economy and Competitiveness (Ref. TIN2012-35874 and TIN2011-28067) and by the Spanish Ministry of Science and Innovation under the cvREMOD CENIT Project (CEN20091044). Authors acknowledge the support from Xavier Planes (UPF, Barcelona, Spain), Marta Sitges (HCPB, Barcelona, Spain) and Alejandro F. Frangi (U of Sheffield, UK).

References

[1] Bordachar, P., Labrousse, L., Ploux, S., Thambo, J.B., Lafitte, S., Reant, P., Jais, P., Haissaguerre, M., Clementy, J., Dos Santos, P.: Validation of a new noninvasive device for the monitoring of peak endocardial acceleration in pigs: implications for optimization of pacing site and configuration. Journal of Cardiovascular Electrophysiology 19(7), 725–729 (2008)

[2] Halvorsen, P.S., Remme, E.W., Espinoza, A., Skulstad, H., Lundblad, R., Bergsland, J., Hoff, L., Imenes, K., Edvardsen, T., Elle, O.J., Fosse, E.: Automatic real-time detection of myocardial ischemia by epicardial accelerometer. J. Thorac. Cardiovasc. Surg. 139(4), 1026–1032 (2010)

[3] Delnoy, P.P., Marcelli, E., Oudeluttikhuis, H., Nicastia, D., Renesto, F., Cercenelli, L., Plicchi, G.: Validation of a peak endocardial acceleration-based algorithm to optimize cardiac resynchronization: early clinical results. Europace 10(7), 801–808 (2008)

[4] Olsen, N., Mogelvang, R., Jons, C., Fritz-Hansen, T., Sogaard, P.: Predicting response to cardiac resynchronization therapy with cross-correlation analysis of myocardial systolic acceleration: a new approach to echocardiographic dyssynchrony evaluation. Journal of the American Society of Echocardiography 22(6), 657–664 (2009)

[5] Ritter, P., Delnoy, P.P., Padeletti, L., Lunati, M., Naegele, H., Borri-Brunetto, A., Silvestre, J.: A randomized pilot study of optimization of cardiac resynchronization therapy in sinus rhythm patients using a peak endocardial acceleration sensor vs. standard methods. Europace 14(9), 1324–1333 (2012)

[6] Remme, E.W., Hoff, L., Halvorsen, P.S., Naerum, E., Skulstad, H., Fleischer, L.A., Elle, O.J., Fosse, E.: Validation of cardiac accelerometer sensor measurements. Physiological Measurement 30(12), 1429–1444 (2009)

[7] Remme, E., Hoff, L., Halvorsen, P., Opdahl, A., Fosse, E., Elle, O.: Simulation model of cardiac three dimensional accelerometer measurements. Medical Engineering and Physics 34(7), 990–998 (2012)

[8] Pashaei, A., Piella, G., Planes, X., Duchateau, N., de Caralt, T.M., Sitges, M., Frangi, A.F.: Image based cardiac acceleration map using statistical shape and 3d+t myocardial tracking models; in-vitro study on heart phantom. In: SPIE Medical Imaging, February 9-14. SPIE, Florida (2013)

[9] De Craene, M., Piella, G., Camara, O., Duchateau, N., Silva, E., Doltra, A., D'hooge, J., Brugada, J., Sitges, M., Frangi, A.F.: Temporal diffeomorphic free-form deformation: application to motion and strain estimation from 3d echocardiography. Medical Image Analysis 16(2), 427–450 (2012)

[10] van Assen, H.C., Danilouchkine, M.G., Frangi, A.F., Ordas, S., Westenberg, J.J.M., Reiber, J.H.C., Lelieveldt, B.P.F.: SPASM: a 3D-ASM for segmentation of sparse and arbitrarily oriented cardiac MRI data. Medical Image Analysis 10(2), 286–303 (2006)

[11] Byrd, R., Lu, P., Nocedal, J., Zhu, C.: A limited memory algorithm for bound constrained optimization. SIAM Journal on Scientific Computing 16(5), 1190–1208 (1995)

[12] Cerqueira, M., Weissman, N., Dilsizian, V., Jacobs, A., Kaul, S., Laskey, W., Pennell, D., Rumberger, J., Ryan, T., Verani, M., et al.: Standardized myocardial segmentation and nomenclature for tomographic imaging of the heart a statement for healthcare professionals from the cardiac imaging committee of the council on clinical cardiology of the american heart association. Circulation 105(4), 539–542 (2002)

Author Index

Abbou, Amine 42
Abi-Nahed, Julien 244
Al-Ansari, Abdulla 244
Albà, Xènia 196
Albal, Priti G. 83
Ammar, Mohammed 42
Andreu, David 220
Avrahami, Idit 110
Ayache, Nicholas 49, 152, 228

Bai, Fan 57
Barker, Alex 236
Beinart, Roy 253
Berruezo, Antonio 220
Betancur, Julián 24
Bhatia, Kanwal K. 126
Bijnens, Bart 220
Bismuth, Jean 94
Biswas, Labonny 152
Blanck, Oliver 31
Bluemke, David A. 143
Brown, Alistair 94
Bruurmijn, L.C. Mark 212
Burgreen, Greg W. 118

Camara, Oscar 220, 262
Cárdenes, Rubén 220
Chikh, Mohammed Amine 14, 42
Cito, Salvatore 74
Comaniciu, Dorin 162, 236, 253
Cooklin, Michael 126
Cowan, Brett R. 143
Criminisi, Antonio 49
Crystal, Eugene 152

Dakua, Sarada Prasad 244
Daoudi, Abdelaziz 14
Debus, Kristian 94
Delingette, Hervé 228
Dillenseger, Jean-Louis 24
Ding, Jinli 57
Dori, Yoav 253
Duchateau, Nicolas 262
Duits, Remco 212

Duncan, James 162
Dur, Onur 83

ElBaz, Mohammed S.M. 204
Ennis, Daniel B. 135

Fabrèges, Benoit 102
Fernández-Armenta, Juan 220
Filatova, Olena G. 212
Finn, J. Paul 143
Flor, Roey 152
Florack, Luc M.J. 212
Frangi, Alejandro F. 196
Fu, Wenyu 57
Fuster, Andrea 212

Gabrielli, Luigi 262
Gahm, Jin Kyu 135
Georgescu, Bogdan 253
Gerbeau, Jean-Frédéric 102
Ghate, Sudip 152
Gill, Jas 126
Ginks, Matthew 228
Gopal, Sharath 180
Goubergrits, Leonid 65
Grbic, Sasa 188
Gulsun, Mehmet 236

Halperin, Henry R. 253
Herzka, Daniel A. 253
Hoogendoorn, Corné 196
Housden, R. James 126, 171

Ionasec, Razvan Ioan 162, 188
Itu, Lucian 236

Kadish, Alan H. 143
Kamen, Ali 236, 253
Kanik, Jingjing 162
Karim, Rashed 1
Karmonik, Christof 94
Kause, Hanne B. 212
Kertzscher, Ulrich 65
King, Andrew P. 126
Kuehne, Titus 65

Lee, Daniel C.   143
Lekadir, Karim   196
Lelieveldt, Boudewijn P.F.   204
Lima, João A.C.   143
Liu, Youjun   57
Lumsden, Alain B.   94

Ma, YingLiang   126, 171
Mahmoudi, Saïd   14, 42
Mansi, Tommaso   162, 188, 253
Margeta, Ján   49
Markl, Michael   236
McLeod, Kristin   49
Medrano-Gracia, Pau   143
Menon, Prahlad G.   83
Mihalef, Viorel   236
Montidoro, Tyson A.   83

Neumann, Dominik   188
Newbigging, Susan   152

Oduneye, Samuel   152
O'Neill, Mark   126

Pallarés, Jordi   74
Panayiotou, Maria   126
Pant, Sanjay   102
Pashaei, Ali   196, 262
Pereanez, Marco   196
Peters, Jochen   1
Piella, Gemma   262
Pinto, Karen   1
Pop, Mihaela   152

Qiao, Aike   57

Rabbah, Jean-Pierre   188
Rapaka, Saikiran   236, 253
Razavi, Reza   1, 228
Relan, Jatin   228
Ren, Xiaochen   57
Rhode, Kawal S.   1, 126, 171, 228
Riesenkampff, Eugénie   65
Rinaldi, C. Aldo   126, 228
Rueckert, Daniel   171

Saikrishnan, Neelakantan   188
Sandoval, Zulma   24
Schaeffter, Tobias   1
Schaller, Jens   65
Schlaefer, Alexander   31
Scorza, Angelo   236
Sebastian, Rafael   220
Sermesant, Maxime   152, 228
Sharma, Puneet   162, 236
Siefert, Andrew W.   188
Soto-Iglesias, David   220
Stender, Birgit   31
Suinesiaputra, Avan   143

Terzopoulos, Demetri   180
Thompson, David S.   118
Tobon-Gomez, Catalina   1

van Assen, Hans C.   212
van der Geest, Rob J.   204
Varma, Niharika   171
Vernet, Anton   74
Vignon-Clementel, Irene E.   102
Voigt, Ingmar   162, 188

Walters, D. Keith   118
Wang, Bo   31
Wang, Xiao   118
Weese, Juergen   1
Westenberg, Jos J.M.   204, 212
Wong, Ken C.L.   228
Wright, Graham A.   152
Wu, Xianliang   171

Yevtushenko, Pavlo   65
Yoganathan, Ajit P.   188
Young, Alistair A.   143
Yuh, David D.   188

Zettinig, Oliver   253
Zhang, Mingzi   57
Zhao, Xi   57
Zviman, M. Muz   253